Springer Theses

Recognizing Outstanding Ph.D. Research

Aims and Scope

The series "Springer Theses" brings together a selection of the very best Ph.D. theses from around the world and across the physical sciences. Nominated and endorsed by two recognized specialists, each published volume has been selected for its scientific excellence and the high impact of its contents for the pertinent field of research. For greater accessibility to non-specialists, the published versions include an extended introduction, as well as a foreword by the student's supervisor explaining the special relevance of the work for the field. As a whole, the series will provide a valuable resource both for newcomers to the research fields described, and for other scientists seeking detailed background information on special questions. Finally, it provides an accredited documentation of the valuable contributions made by today's younger generation of scientists.

Theses are accepted into the series by invited nomination only and must fulfill all of the following criteria

- They must be written in good English.
- The topic should fall within the confines of Chemistry, Physics, Earth Sciences, Engineering and related interdisciplinary fields such as Materials, Nanoscience, Chemical Engineering, Complex Systems and Biophysics.
- The work reported in the thesis must represent a significant scientific advance.
- If the thesis includes previously published material, permission to reproduce this must be gained from the respective copyright holder.
- They must have been examined and passed during the 12 months prior to nomination.
- Each thesis should include a foreword by the supervisor outlining the significance of its content.
- The theses should have a clearly defined structure including an introduction accessible to scientists not expert in that particular field.

More information about this series at http://www.springer.com/series/8790

Daisuke Ogura

Theoretical Study of Electron Correlation Driven Superconductivity in Systems with Coexisting Wide and Narrow Bands

Doctoral Thesis accepted by
Osaka University, Osaka, Japan

Author
Dr. Daisuke Ogura
Department of Physics
Osaka University
Osaka, Japan

Supervisor
Prof. Kazuhiko Kuroki
Department of Physics
Graduate School of Science
Osaka University
Osaka, Japan

ISSN 2190-5053　　　　　　　ISSN 2190-5061　(electronic)
Springer Theses
ISBN 978-981-15-0666-6　　　ISBN 978-981-15-0667-3　(eBook)
https://doi.org/10.1007/978-981-15-0667-3

© Springer Nature Singapore Pte Ltd. 2019
This work is subject to copyright. All rights are reserved by the Publisher, whether the whole or part of the material is concerned, specifically the rights of translation, reprinting, reuse of illustrations, recitation, broadcasting, reproduction on microfilms or in any other physical way, and transmission or information storage and retrieval, electronic adaptation, computer software, or by similar or dissimilar methodology now known or hereafter developed.
The use of general descriptive names, registered names, trademarks, service marks, etc. in this publication does not imply, even in the absence of a specific statement, that such names are exempt from the relevant protective laws and regulations and therefore free for general use.
The publisher, the authors and the editors are safe to assume that the advice and information in this book are believed to be true and accurate at the date of publication. Neither the publisher nor the authors or the editors give a warranty, expressed or implied, with respect to the material contained herein or for any errors or omissions that may have been made. The publisher remains neutral with regard to jurisdictional claims in published maps and institutional affiliations.

This Springer imprint is published by the registered company Springer Nature Singapore Pte Ltd.
The registered company address is: 152 Beach Road, #21-01/04 Gateway East, Singapore 189721, Singapore

Supervisor's Foreword

Seeking for high-temperature superconductivity is one of the most challenging issues in the field of condensed matter physics. In conventional superconductors, Cooper pairs are formed by the attractive pairing interaction between electrons, which is mediated by phonons. In this case, however, the typically low energy of the phonons tend to suppress the superconducting transition temperature (T_c). In this sense, one may expect higher T_c from purely electronic, and thus unconventional, pairing mechanisms. In fact, it has been shown in some previous studies that superconductivity with very high T_c may be realized in purely electronic systems with coexisting wide and narrow bands, provided that the Fermi level lies in the vicinity of the narrow band edge. However, a systematic understanding on the role of strong electron correlation effects in such a pairing mechanism is yet to be attained, and also a realistic candidate material which meets the ideal condition is unknown.

Given the above background, Daisuke, in the former part of his thesis, applied FLEX+DMFT method to multiband systems, where all the coding and calculation were done by himself. This is a method combining the fluctuation exchange (FLEX) approximation, which treats non-local correlations perturbatively, and the dynamical mean-field theory (DMFT), which describes local correlations non-perturbatively. By performing systematic calculations on systems with coexisting wide and narrow bands, Daisuke showed that the above mentioned multiband pairing mechanism is indeed valid even when the strong correlation effect is taken into account within the FLEX+DMFT formalism. In the latter part of the thesis, Daisuke proposed realistic candidates for realizing the ideal situation for high T_c superconductivity. The key concept here is the "hidden ladder" electronic structure, which is realized due to the combination of the bilayer lattice structure and the anisotropy of the orbitals of the electrons. The achievement of Daisuke's thesis can be viewed as giving a theoretical guiding principle for seeking unknown high T_c superconductors.

Osaka, Japan
May 2019

Prof. Kazuhiko Kuroki

List of Publications

Parts of this thesis have been published in the following journal articles:

- Daisuke Ogura, Hideo Aoki, and Kazuhiko Kuroki, "Possible high-Tc superconductivity due to incipient narrow bands originating from hidden ladders in Ruddlesden-Popper compounds", Physical Review B **96**, 184513 (2017), part of Chapter 5.
- Masahiro Nakata, Daisuke Ogura, Hidetomo Usui, and Kazuhiko Kuroki, "Finite energy spin fluctuations as a pairing glue in systems with coexisting electron and hole bands", Physical Review B **95**, 214509 (2017).
- Karin Matsumoto, Daisuke Ogura, and Kazuhiko Kuroki, "Wide applicability of high-Tc pairing originating from coexisting wide and incipient narrow bands in quasi-one-dimensional systems", Physical Review B **97**, 014516 (2018).

Current Affiliation of the Author and His Supervisor

Author: Dr. Daisuke Ogura, Hitachi Ltd., Japan
Supervisor: Prof. Kazuhiko Kuroki, Osaka University, Osaka, Japan

Acknowledgements

First and foremost, I would like to express my gratitude to my advisor, Prof. Kazuhiko Kuroki, who provided helpful comments and suggestions throughout my Ph.D. course. His constant support during my thesis work was indispensable to complete the present thesis. I would like to thank Prof. Yasuhiro Akutsu, Prof. Tamio Oguchi, Prof. Keith Slevin, and Prof. Shigeki Miyasaka for being members of my committee.

I am very grateful to Prof. Hideo Aoki for the collaboration in the study of the hidden ladders. I appreciate Hiroshi Eisaki, Kenji Kawashima, Shigeyuki Ishida, Hiraku Ogino, and Yoshiyuki Yoshida for illuminating discussions on realization of the hidden ladders in actual materials. I also thank Masayuki Ochi for fruitful discussions and useful comments, and Katsuhiro Suzuki for assistance with the multi-orbital FLEX code, and all the members of the condensed matter theory groups for their various helps.

Last but not least, I would like to thank my family for giving me constant encouragements and supports through my life.

Contents

1 Introduction 1
   1.1 Superconductivity 1
      1.1.1 Discovery of Superconductivity 1
      1.1.2 Conventional/Unconventional Superconductivity 1
   1.2 Cuprate Superconductors 2
      1.2.1 Discovery of Cuprate Superconductors 3
      1.2.2 Properties of Cuprate Superconductors 3
   1.3 Iron-Based Superconductors 5
      1.3.1 Discovery of Iron-Based Superconductors 5
      1.3.2 Properties of Iron-Based Superconductors 5
   References 9

2 Background of the Present Thesis 11
   2.1 Superconductivity in Ladder Systems 11
      2.1.1 Superconductivity in Ladder and Bilayer Models 11
      2.1.2 Experiment on Two-Leg Ladder Compounds 14
   2.2 Superconductivity in Systems with Coexisting Wide
      and Narrow Bands 15
      2.2.1 An Idea for Enhancing T_c 15
      2.2.2 Previous Studies on Systems with Coexisting Wide
            and Narrow Bands 16
   2.3 FLEX + DMFT Method for Single-Band Systems 21
   References 22

3 Method 25
   3.1 Tight-Binding Model 25
      3.1.1 Wannier Functions 25
      3.1.2 Tight-Binding Model 26

	3.2	Green's Function Method for Many-Body Problem	28
		3.2.1 Green's Function	28
		3.2.2 Lehmann Representation and Retarded Green's Function	29
		3.2.3 Thermodynamic Potential	30
		3.2.4 Feynman Diagram	32
		3.2.5 Perturbation Theory for Green's Function	33
		3.2.6 Luttinger–Ward Functional	36
		3.2.7 BCS Theory	37
		3.2.8 Eliashberg Equation	39
	3.3	Perturbative Approximations for Green's Function	42
		3.3.1 Spin/Charge Susceptibility	42
		3.3.2 Random Phase Approximation (RPA)	43
		3.3.3 Fluctuation Exchange (FLEX) Approximation	45
	3.4	Dynamical Mean-Field Theory and Its Combination with FLEX	47
		3.4.1 Path-Integral Representation	47
		3.4.2 Dynamical Mean-Field Theory (DMFT)	48
		3.4.3 Impurity Solvers	53
		3.4.4 FLEX+DMFT Method	56
	3.5	Density Functional Theory	60
		3.5.1 Born-Oppenheimer Approximation	60
		3.5.2 Hohenberg-Kohn Theorem	61
		3.5.3 Kohn-Sham Equation	63
		3.5.4 Exchange-Correlation Functional	65
		3.5.5 Pseudopotential Method	68
		3.5.6 Augmented Plane Wave Method	69
	3.6	Model Construction from DFT Results	71
		3.6.1 Maximally-Localized Wannier Functions	72
	References		73
4	**FLEX + DMFT Analysis for Superconductivity in Systems with Coexisting Wide and Narrow Bands**		**75**
	4.1	Motivation	75
	4.2	Formulation	76
		4.2.1 Model	76
		4.2.2 Many-Body Analysis	78
	4.3	Results and Discussion	80
		4.3.1 Superconductivity	80
		4.3.2 Spectral Function	85
	4.4	Summary	90
	References		90

5 Possible High-T_c Superconductivity in "Hidden Ladder" Materials 91
  5.1 Motivation 91
  5.2 Formulation 92
    5.2.1 Intuitive Idea 92
    5.2.2 Band Calculation and Many-body Analysis 93
  5.3 Results 94
    5.3.1 Band Structure and Effective Model 94
    5.3.2 Many-body Analysis 98
  5.4 Discussion 104
    5.4.1 Relation to Experimental Observations
          in Actual Materials 104
    5.4.2 Other Candidates for Hidden Ladder 107
  5.5 Summary 110
  References 110

6 Conclusion 113

Appendix A: Numerical Analytic Continuation 115

Appendix B: Convergence of the FLEX+DMFT Results Against Computational Parameters 119

Appendix C: Comparison of Spectra Obtained with the Padé Approximation and the Maximum Entropy Method 121

List of Figures

Fig. 1.1	A timeline of the superconducting transition temperature of major families of superconductors. Black points indicate the conventional BCS-type superconductors, red the cuprates, blue the iron-based, and green the organic	3
Fig. 1.2	Typical crystal structures of the cuprate superconductors. The panel (**a**) represents the structure of La_2CuO_4, the panel (**b**) that of $Tl_2Ba_2CaCu_2O_8$, and the panel (**c**) that of Nd_2CuO_4. Blue spheres represent copper atoms, and red oxygen atoms. To plot the crystal structure, we use the VESTA software [11] throughout the present thesis.	4
Fig. 1.3	Typical crystal structures of the iron-based superconductors. Black lines denote the unit cell. Ocher spheres represent iron atoms and green spheres pnictogen/chalcogen atoms. The panel (**a**) represents the structure of the 1111 systems (LaFeAsO), the panel (**b**) that of the 11 systems (FeSe), and the panel (**c**) that of the 122 system (KFe_2As_2)	6
Fig. 1.4	**a** A top view of the FeAs layer, where $As_{above(below)}$ indicating As above (below) Fe plane. Green and red square indicate the single-Fe and original unit cell. **b** A schematic of the unfolded and original (folded) Brillouin zone (Color figure online).	7
Fig. 1.5	**a** A schematic of the 2D Fermi surface of the five-orbital model. In the panel **b**, a cartoon of $s_{\pm}$-wave gap and nesting vectors are depicted	8
Fig. 1.6	Band structure obtained from (color plot) the ARPES measurement and (red lines) the first principles calculation. In the ARPES band structure, it can be seen that the hole Fermi surfaces around the Γ point (the origin of the Brillouin zone) are missing. Reprinted with permission from Ref. [52], Copyright 2011 by the American Physical Society	8

Fig. 1.7	Color plot for the superconducting gap function against the temperature T and the position of the hole band maximum E_h measured from the chemical potential, namely, $E_h < 0$ corresponds to the incipient band case. Gray shaded area indicates the magnetic instability. Reprinted with permission from Ref. [62], Copyright 2016 by the American Physical Society	8
Fig. 2.1	Schematics of lattice structures of **a** a ladder and **b** a bilayer systems	12
Fig. 2.2	Spin gap as a function of the ratio J'/J for different lattice sizes, where $J^{(\prime)}$ is the exchange interaction along a leg (rung). Filled (open) symbols indicate the results for the undoped (hole-doped) case. Reprinted with permission from Ref. [1], Copyright 1992 by the American Physical Society	13
Fig. 2.3	Superconducting correlation as a function of the distance for various strengths of J'. Open triangles, filled triangles, open squares, open hexagons, and filled squares correspond to the results for $J'/J = 0.025, 1, 4, 10$, and 100, respectively. Reprinted with permission from Ref. [1], Copyright 1992 by the American Physical Society	13
Fig. 2.4	A schematic of coexisting wide and incipient narrow bands	16
Fig. 2.5	Carrier concentration dependence of T_c for the Hubbard model on a two leg ladder with diagonal hopping t'. t_l denotes the hopping integral along a leg of a ladder, which is typically ~ 0.4 eV in the ladder cuprate, and n denotes the number of electrons per site. Reprinted with permission from Ref. [27], Copyright 2005 by the American Physical Society	17
Fig. 2.6	**a** Lattice and **b** non-interacting band structures of the diamond chain model. t' denotes the interapex hopping integral, and for $t' = 0(1)$, the middle (bottom) band becomes perfectly flat. Reprinted with permission from Ref. [28], Copyright 2016 by the American Physical Society	17
Fig. 2.7	Binding energy of Cooper pairs. The top panel depicts the color plot against n and $t'^p r$, where t' is the hopping shown in Fig. 2.6 and n is the number of electrons per site per spin and (a–d) some cuts of that along a constant n or $t'^p r$. Reprinted with permission from Ref. [28], Copyright 2016 by the American Physical Society	18
Fig. 2.8	Superconducting correlation function against real space distance for various pair structures shown in the panel (**a**). D and S are the density and spin correlation functions, respectively. Reprinted with permission from Ref. [28], Copyright 2016 by the American Physical Society	19

Fig. 2.9	Band filling dependence of T_c obtained with (left panel) the FLEX approximation and (right) FLEX + DMFT method. Filling is defined as the number of electrons per unit cell and filling = 1 corresponds to the half-filled case. The panel (**a**) is result for the single-band Hubbard model on a square lattice only with the nearest neighbor hopping, and the panel (**b**) is for the single-band Hubbard model on a square lattice with the next nearest neighbor and the third neighbor hoppings. Reprinted with permission from Ref. [33], Copyright 2015 by the American Physical Society.	20
Fig. 2.10	Spectral functions of the single-band Hubbard model on a square lattice with the next nearest neighbor and the third neighbor hoppings for some carrier concentrations n obtained with the FLEX + DMFT method. At half-filling ($n = 1$), there can be seen a dip structure at low energy. Reprinted with permission from Ref. [33], Copyright 2015 by the American Physical Society	21
Fig. 2.11	Fermi surfaces of the single-band Hubbard model (left) above and (right) below the Pomeranchuk instability temperature T_c^{PI} obtained with the FLEX + DMFT method. Reprinted with permission from Ref. [34], Copyright 2017 by the American Physical Society	21
Fig. 3.1	Diagrammatic representation of the first order contribution $\Omega^{(1)}$.	33
Fig. 3.2	First-order disconnected diagrams.	34
Fig. 3.3	First-order connected diagrams	35
Fig. 3.4	A diagrammatic representation of the Dyson equation	36
Fig. 3.5	Diagrammatic expression of the Eliashberg equation. Double-headed arrow means the anomalous Green's function.	41
Fig. 3.6	Feynman diagrams of the susceptibilities within RPA for the single-orbital Hubbard model	45
Fig. 3.7	Example for bubble- and ladder-type diagrams of the Luttinger–Ward functional considered in the FLEX approximation.	46
Fig. 3.8	A flowchart for the procedure to solve the Kohn–Sham equation	66
Fig. 4.1	**a** A schematic of the bilayer model. Arrows indicate hopping integrals. **b** A schematic of the anisotropic bilayer model, where the dashed lines represent a weak coupling toward the y direction. **c** A schematic of the dimer array model, where the thick lines represent a strong intra-dimer coupling	77

Fig. 4.2	Band structures for **a** the wide-and-flat-band model, **b** the wide-and-narrow-band model, **c** the anisotropic bilayer model, and **d** the dimer-array model in which we take $t_d/t = 2$.	79
Fig. 4.3	Non-interacting density of states for **a** the wide-and-flat-band model, **b** the wide-and-narrow-band model, **c** the anisotropic bilayer model, and **d** the dimer-array model in which we take $t_d/t = 2$.	80
Fig. 4.4	(Left panel) FLEX+DMFT results of the band filling dependence of (upper panel) the eigenvalue λ of the linearized Eliashberg equation and (lower panel) the Stoner factor for the wide-and-flat-band model. For comparison, the FLEX results are also displayed in the right panels	81
Fig. 4.5	(Left panel) FLEX+DMFT results of the band filling dependence of (upper panel) the eigenvalue λ of the linearized Eliashberg equation and (lower panel) the Stoner factor for the wide-and-narrow-band model. For comparison, the FLEX results are also displayed in the right panels.	82
Fig. 4.6	(Left panel) FLEX+DMFT results of the band filling dependence of (upper panel) the eigenvalue λ of the linearized Eliashberg equation and (lower panel) the Stoner factor for the anisotropic bilayer model. For comparison, the FLEX results are also displayed in the right panels	82
Fig. 4.7	(Left panel) FLEX+DMFT results of the interlayer coupling strength dependence of (upper panel) the eigenvalue λ of the linearized Eliashberg equation and (lower panel) the Stoner factor for the dimer array model. For comparison, the FLEX results are also displayed in the right panels.	83
Fig. 4.8	Anomalous self-energy for the wide-and-flat-band model at the optimal filling $n = 2.26$. Panels **a** and **b** show the intralayer component Δ_{11} and **b** the interlayer component Δ_{12}. In lower panels, the anomalous self-energy in the band representation **c** for the band 1 $\Delta_{11} + \Delta_{12}$ and **d** for the band 2 $\Delta_{11}-\Delta_{12}$	84
Fig. 4.9	Anomalous self-energy for the wide-and-narrow-band model at the optimal filling $n = 2.36$. Panels **a** and **b** show the intralayer component Δ_{11} and **b** the interlayer component Δ_{12}. In lower panels, the anomalous self-energy in the band representation **c** for the band 1 $\Delta_{11} + \Delta_{12}$ and **d** for the band 2 $\Delta_{11}-\Delta_{12}$	85
Fig. 4.10	Anomalous self-energy for the anisotropic bilayer model at the optimal filling $n = 1.32$. Panels **a** and **b** show the intralayer component Δ_{11} and **b** the interlayer component Δ_{12}. In lower panels, the anomalous self-energy in the band representation **c** for the band 1 $\Delta_{11} + \Delta_{12}$ and **d** for the band 2 $\Delta_{11}-\Delta_{12}$	86

Fig. 4.11	Anomalous self-energy for the dimer array model at the optimal coupling strength $t_d/t = 1.75$. Panels **a** and **b** show the intralayer component Δ_{11} and **b** the interlayer component Δ_{12}. In lower panels, the anomalous self-energy in the band representation **c** for the band 1 $\Delta_{11} + \Delta_{12}$ and **d** for the band 2 $\Delta_{11} - \Delta_{12}$	87
Fig. 4.12	(Left panels) FLEX + DMFT result of the spectral function of the wide-and-narrow-band model for (a–1) a too small band filling, (a–2) the optimal band filling, and (a–3) a too large band filling. For comparison, the FLEX result is also displayed in the right panels (b–1–3). The vertical dashed line represents the chemical potential	88
Fig. 4.13	(Left panels) FLEX + DMFT result of the spectral function of the anisotropic bilayer model for (a–1) a too small band filling, (a–2) the optimal band filling, and (a–3) a too large band filling. For comparison, the FLEX result is also displayed in the right panels (b–1–3). The vertical dashed line represents the chemical potential	89
Fig. 5.1	Schematics of "hidden ladders" composed of the d_{xz} (left panel) and d_{yz} (right) orbitals in the bilayer Ruddlesden-Popper compounds $Sr_3TM_2O_7$ (*TM* indicates a transition metal). The middle panel depicts the crystal structure	92
Fig. 5.2	Band structures of **a, b** $Sr_3Mo_2O_7$ for the experimental structure, **c, d** $Sr_3Mo_2O_7$ for the optimized structure, and **e, f** $Sr_3Cr_2O_7$ obtained from the first-principles calculation. The Brillouin zone is shown in the inset. The size of the blue circles in the left (right) panels depicts the weight of the $d_{xy}(d_{xz,yz})$ orbital character (Color figure online)	95
Fig. 5.3	Density of states of **a** $Sr_3Mo_2O_7$ for the experimental structure, **b** $Sr_3Mo_2O_7$ for the optimized structure, and **c** $Sr_3Cr_2O_7$ obtained from the first-principles calculation. Purple lines denote the total density of states along with the projected density of states for the *TM* atom, the *TM* d_{xy} orbital, and the *TM* d_{xz+yz} orbital by green, orange, and cyan lines, respectively. Dashed vertical lines represents the Fermi level	96
Fig. 5.4	The band structure of a ladder cuprate $SrCu_2O_3$ is displayed along two paths in the conventional Brillouin zone, where the thickness of the lines represents the $d_{x^2-y^2}$ orbital character. The point $(\pi, \pi, 0)$ corresponds to Γ point in Fig. 5.2 because the sign of the hopping is reversed. Note that	

	panels (**a**) and (**b**) superimposed form a band structure quite similar to those in Fig. 5.2b, d, respectively, except for the position of the Fermi level	97
Fig. 5.5	Band structure (red lines) of the six-orbital tight-binding model for $Sr_3Mo_2O_7$ constructed from the maximally-localized Wannier orbitals. Gray lines represent the first-principles calculation.	97
Fig. 5.6	FLEX result for the band filling dependence of **a** the eigenvalue λ of the linearized Eliashberg equation and **b** the Stoner factor for $Sr_3Mo_2O_7$ for $U = 2.5$ eV (red diamonds), $U = 3.0$ eV (purple diamonds) for the experimental structure and for $U = 2.5$ eV for the optimized structure. The vertical dashed line represents the stoichiometric point ($n = 4$). Also displayed are the results for the 326 compound $Sr_3Mo_2O_6$ (orange triangle) and a F-doped $Sr_3Mo_2O_6F$ (blue inverted triangle) at $n = 4$ (Color figure online).	99
Fig. 5.7	Obtained anomalous self-energy for **a** $n = 4.2$, **b** $n = 3.8$ (optimial filling for superconductivity), and **c** $n = 4.2$. The subscripts $\{0, 1, 2, 3, 4, 5\}$ denote the $\{d^1_{xy}, d^1_{xz}, d^1_{yz}, d^2_{xy}, d^2_{xz}, d^2_{yz}\}$ orbitals, with the superscript 1 or 2 indicating the layer index	100
Fig. 5.8	Anomalous self-energy for **a** $n = 4.2$, **b** $n = 3.8$ (optimal filling for superconductivity), and **c** $n = 4.2$ in the band representation (black dots) obtained with the unitary transformation which diagonalizes the six-orbital tight-binding Hamiltonian. For intuitive understanding, linear combinations of the intra- and inter-layer component for the $d_{xy}/d_{xz}/d_{yz}$ orbital are also shown. Purple, green, and cyan lines correspond to the bonding orbitals comprising the d_{xy}, d_{xz}, and d_{yz} orbital, respectively, and, orange, yellow, and blue lines to the anti-bonding orbitals.	101
Fig. 5.9	Renormalized density of states obtained with the Padé approximation. Left panels show the projected density of states of the $d_{xz,yz}$ (purple lines) and d_{xy} orbitals (green) along with the total density of states (black). Right panels show the projected density of states of $d_{xz,yz}/d_{xy}$ decomposed into the bonding (purple/cyan) and anti-bonding orbitals (green/orange). The vertical dashed line represents the chemical potential.	103
Fig. 5.10	Renormalized density of states obtained with the Padé approximation for $n = 2.8$. Panels are similar to Fig. 5.9. The vertical dashed line represents the chemical potential	104

Fig. 5.11	Band structures of **a**, **b** $Sr_3Mo_2O_6F$, **c**, **d** $Sr_3Mo_2O_6$. The size of the blue circles in the left (right) panels depicts the weight of the $d_{xy}(d_{xz,yz})$ orbital character (Color figure online)	106
Fig. 5.12	Band structures of fluorine-intercalated compounds **a** $La_2SrCr_2O_7F_2$ and **b** $La_2SrMo_2O_7F_2$	107
Fig. 5.13	A schematic hidden ladder in the triple-layer Ruddlesden-Popper materials, here displayed for d_{yz} orbitals	108
Fig. 5.14	Band structures of triple-layer compounds **a**, **b** $Sr_4Cr_3O_{10}$ and **c**, **d** $La_4Cr_3O_{10}$. The thickness of the lines in the left(right) panels depicts the weight of the $d_{xy}(d_{xz,yz})$ orbital character. **e**, **f** For comparison, the band structure of a three-leg ladder cuprate $Sr_2Cu_3O_5$ is shown along two paths in the conventional Brillouin zone, where the thickness of the lines represents the $d_{x^2-y^2}$ character. Note that the sign of the energy is reversed to facilitate comparison between the t_{2g} (present) and e_g (cuprate) systems. As in the two-layer case, panels (e–1) and (e–2) [(f–1) and (f–2)] superimposed form a band structure quite similar to those in (a–2) and (c–2) [(b–2) and (d–2)]. Red dashed lines indicate the edge of the narrow bands (Color figure online)	109
Fig. B.1	The eigenvalue λ of the linearized Eliashberg equation for some sets of computational parameters	119
Fig. B.2	The band filling dependence of the eigenvalue λ of the linearized Eliashberg equation for $n_{Mat} = 2048$. For comparison, the same plot for $n_{Mat} = 1024$ is also shown	120
Fig. C.1	Spectral functions for the bonding/antibonding bands in the anisotropic bilayer model. Upper panel depicts those obtained with the maximum entropy method, and lower panel those obtained with the Padé approximation	121

Chapter 1
Introduction

Abstract In this chapter, we will introduce some basics of superconductivity related to the present study. Firstly, we briefly mention superconductivity and classification of it into conventional and unconventional superconductivity. We also describe some basics of two material classes of high superconducting transition temperatures, the cuprate and iron-based superconductors, which are considered to be typical families of unconventional superconductors.

Keywords Superconductivity · Cuprate superconductors · Iron-based superconductors.

1.1 Superconductivity

1.1.1 Discovery of Superconductivity

Superconductivity is a phenomenon emerging below a certain temperature T_c (superconducting transition temperature), which is characterized by perfect conductivity and perfect diamagnetism (Meissner effect). It was discovered by Onnes [1] in solid mercury with $T_c = 4.2$ K.

1.1.2 Conventional/Unconventional Superconductivity

According to the BCS theory, which we will describe in Chap. 3, superconductivity emerges by obtaining energy gain (superconducting gap) due to forming electron pairs (Cooper pairs) driven by a phonon-mediated attractive interaction between electrons. The above-mentioned features such as perfect conductivity and perfect diamagnetism can be naturally understood within the BCS theory. As one of the most important experimental verifications of the BCS theory, the isotope effect on T_c is well-known. In the case of phonon-mediated superconductivity described by the

BCS theory, since T_c is characterized by the Debye frequency, the relation between T_c and the atomic mass M becomes $T_c \propto M^{-1/2}$. This relation was confirmed experimentally by the isotope effect in almost all elemental superconductors and hence is considered to be one of the strongest evidences for the phonon-mediated pairing mechanism. Within this mechanism, T_c is roughly determined by the Debye frequency (typically $\sim \mathcal{O}(100)$ K), which characterizes the energy scale of phonons. Based on the strong coupling theory for the phonon-mediated superconductivity, which we will describe in Chap. 3, the maximum T_c within this mechanism can be roughly estimated as about 10% of the Debye frequency. Therefore it is considered to be difficult to have superconductivity with T_c exceeding 100 K within this mechanism, except for some examples in extremely high pressure such as the sulfur hydride ($T_c \sim 203$ K at ~ 150 GPa) [2] and the lanthanum hydride ($T_c \sim 215$ K at ~ 150 GPa in Ref. [3] and $T_c \sim 260$ K at ~ 200 GPa in Ref. [4]), reported recently.

While phonon-mediated effective interaction is considered as the driving force of superconductivity in the BCS theory [5], it is possible to consider electron correlation driven superconductivity in general. For instance, the cuprates and the iron-based superconductors, which we will describe later, are considered to be typical examples of superconductors based on an electron-correlation-driven pairing mechanism, e.g. a spin-fluctuation-mediated pairing. For these reasons, phonon-mediated and electron correlation driven superconductivity are called conventional and unconventional superconductivities, respectively. In unconventional superconductors, since the original energy scale of the electrons is $\sim \mathcal{O}(1)$ eV, high-T_c's can be expected.

In the BCS theory, a constant superconducting gap is assumed. However, in general, the superconducting gap can be anisotropic and/or change its sign in the Brillouin zone. For instance, let us consider an electron correlation driven pairing mechanism with a repulsive pairing interaction characterized by a certain momentum transfer such as a spin fluctuation mediated pairing mechanism. In this case, the superconducting gap has to change its sign in the Brillouin zone against the characteristic momentum transfer of the pairing interaction. Since the anisotropy of the superconducting gap is characterized by the angular momentum l of the pair wave function, superconductivity with $l = 0, 1, 2, \ldots$ is called $s, p, d, \ldots$-wave superconductivity. Zeros of the superconducting gap along certain directions in the momentum space are called nodes, which play a key role in determining the symmetry of the superconducting gap.

1.2 Cuprate Superconductors

In this section, we briefly review some basic properties of the cuprate superconductors. For more details, we refer to Refs. [6, 7].

1.2 Cuprate Superconductors

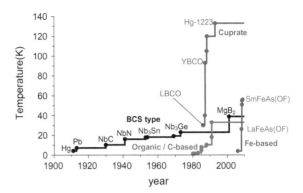

Fig. 1.1 A timeline of the superconducting transition temperature of major families of superconductors. Black points indicate the conventional BCS-type superconductors, red the cuprates, blue the iron-based, and green the organic

1.2.1 Discovery of Cuprate Superconductors

Superconductivity in cuprates was first reported by Bednorz and Müller in 1986 [8]. It was in the non-stoichiometric $La_{2-x}Ba_xCuO_4$ and its T_c is about 30 K, which was record-breaking. Subsequently, many high-T_c cuprate superconductors were discovered by highly extensive studies, as exampled by superconductivity with T_c of about 90 K in $YBa_2Cu_3O_{7-\delta}$ [9]. This fact strongly suggests that superconductivity observed in this series of cuprates is not a conventional one. In Fig. 1.1, we show a timeline of the superconducting transition temperature of major families of superconductors. At present, the highest T_c at ambient pressure among known compounds is $T_c \sim 133$ K of a mercury-based cuprate $HgBa_2Ca_2Cu_3O_8$ [10].

1.2.2 Properties of Cuprate Superconductors

The cuprates having high T_c's, say $T_c > 30$ K, have layered structure composed of conducting CuO_2 layers and insulating blocking layers as shown in Fig. 1.2, which makes these materials quasi-two-dimensional systems. A common feature among the high-T_c cuprates is a CuO_2 layer that can be regarded as a two dimensional square lattice of Cu-O (elongated) octahedra, pyramids, or squares. Another important feature is that the valence of the known high-T_c cuprates is Cu^{2+} in the mother (undoped) compounds, namely the electron configuration of a Cu atom is $[Ar]3d^94s^0$. This means only the $d_{x^2-y^2}$ orbital (more precisely $Cud_{x^2-y^2}$-Op_σ anti-bonding orbital) contribute to the conduction, and the $d_{x^2-y^2}$ orbital is half-filled.

Usually, the mother compounds of the high-T_c cuprate are known to be antiferromagnetic insulators in the stoichiometry corresponding to Cu^{2+}, which strongly suggests that the insulating state in the cuprate is driven by strong electron correlations, and superconductivity occurs by doping electrons or holes. Given this property, the mechanism for high-T_c superconductivity in this material class has been theoretically studied using models such as the Hubbard model (a tight-binding model with

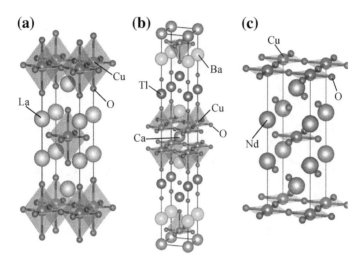

Fig. 1.2 Typical crystal structures of the cuprate superconductors. The panel (**a**) represents the structure of La_2CuO_4, the panel (**b**) that of $Tl_2Ba_2CaCu_2O_8$, and the panel (**c**) that of Nd_2CuO_4. Blue spheres represent copper atoms, and red oxygen atoms. To plot the crystal structure, we use the VESTA software [11] throughout the present thesis

the on-site repulsion, see Chap. 3), or the t-J model (a model obtained by introducing holes in the Heisenberg model) have been extensively studied on a square lattice. The pairing mechanism is likely to be related to antiferromagnetic spin fluctuaions, but a consensus regarding the pairing mechanism has not been reached yet.

In the hole-doped cuprates, it is well-known that T_c exhibits a dome-like doping dependence against hole doping rate (e.g. Ref. [12]). Namely, for a small doping rate (under-doped region), T_c increases with doping and takes maximum around 15% doping (optimal doping). For further hole doping (over-doped region), T_c turns to decrease. This feature does not depend on the blocking layers or the number of conducting layers [13] in many cases [14].

In the under-doped region, a finite excitation energy gap is observed even above T_c, which is called the pseudogap phase [15, 16]. This was firstly observed with the nuclear magnetic resonance (NMR) [17]. Subsequently, the pseudogap has been observed by many experimental probes such as angle resolved photoemission spectroscopy (ARPES) [18, 19], electrical conductivity [20, 21], thermoelectric power [22], and optical conductivity [23]. Therefore there is an excitation gap not only in the spin but also in the charge excitation. There are also experiments indicating multiple gaps rather than single [24–26], which are distinguished from each other by the size of the pseudogap. There are also reports on a charge density wave in the pseudogap phase, e.g. Ref. [27] which has also attracted attention. As for the origin of the pseudogap, precursor of the Mott transition, that of the antiferromagnetic transition, that of superconducting transition, and so on are considered as candidates but consensus has not been reached. In any case, the pseudogap behavior in strongly correlated electron systems is considered to be a hallmark of the strong electron correlation effect.

As for the symmetry of the superconducting gap, the high-T_c cuprates are experimentally known to be $d_{x^2-y^2}$-superconductors evidenced by NMR [28–30], penetration depth [31], ARPES [32, 33], scanning tunneling microscopy [34], and interference experiments in the junctions [35–39].

In order to treat this system, the single-band Hubbard model on a square lattice, which is introduced in Chap. 3, is often adopted as a simplest model which captures essential features of the system since it can describe d-wave superconductivity, the strong magnetic fluctuations, and the Mott transition, which is a metal-insulator transition due to strong correlations.

1.3 Iron-Based Superconductors

In this section we briefly review some basic properties of the iron-based superconductors. For more details, we refer to Refs. [40, 41].

1.3.1 Discovery of Iron-Based Superconductors

The first superconductivity in the iron-based compound was found in LaFePO in 2006 [42]. However, because of its low T_c, it attracted less attention at that time. Since the discovery of relatively high-T_c superconductivity of LaFeAsO$_{1-x}$F$_x$ with $T_c = 26$ K [43], it has been studied extensively. The highest T_c among bulk samples of the iron-based superconductors at ambient pressure is about 55 K, which is the highest next to the cuprates. Thanks to the extensive studies, superconductivity in various iron pnictides and chalcogenides has been found so far.

1.3.2 Properties of Iron-Based Superconductors

Similar to the high-T_c cuprates, the iron-based superconductors have layered structure basically composed of conducting FePn (Pn: pnictgen) [or FeCh (Ch: chalcogen)] layers and insulating blocking layers as shown in Fig. 1.3. Note that FeSe and its analogs (Fig. 1.3b) have only conducting layers. In a conducting layer, Fe atoms (with the formal valence of 2+ in the undoped materials), which are tetrahedrally coordinated by Pn or Ch atoms, form a square lattice. The iron-based superconductors are often called by their compositional ratio, e.g. LaFePO and its analogs (Fig. 1.3a) are called as the 1111 systems, FeSe and its analogs (Fig. 1.3b) the 11 system, KFe$_2$As$_2$ its analogs (Fig. 1.3c) the 122 systems, and so on.

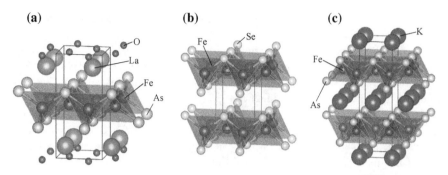

Fig. 1.3 Typical crystal structures of the iron-based superconductors. Black lines denote the unit cell. Ocher spheres represent iron atoms and green spheres pnictogen/chalcogen atoms. The panel (**a**) represents the structure of the 1111 systems (LaFeAsO), the panel (**b**) that of the 11 systems (FeSe), and the panel (**c**) that of the 122 system (KFe$_2$As$_2$)

Unlike in the cuprates, the electronic structure near the Fermi level is somewhat complicated in the iron-based superconductors. According to first principles calculations, there are 10 bands mainly comprising the Fe d-orbitals near the Fermi level since the unit cell contains two tetrahedra composed of, say FeAs. However, focusing only on Fe atoms, the unit cell can be reduced so that the unit cell contains only one Fe atom (single-Fe unit cell) in some, but not all cases. Therefore the electronic structure near the Fermi level can be "unfolded" into a Brillouin zone which is twice larger than that of the original one, as schematically shown in Fig. 1.4. Namely a minimal, but realistic model of the iron-based superconductors comprises the five d-orbitals. The Fermi surface of the five-orbital model is schematically shown in Fig. 1.5a. We call two Fermi surfaces around $(0, 0)$, which are known to be hole Fermi surfaces and originating from the $d_{XZ/YZ}$ orbitals, as α_1 and α_2, and two around $(\pi, 0)$ and $(0, \pi)$, which are known to be electron Fermi surfaces and originating from the $d_{XZ/YZ}$ and $d_{X^2-Y^2}$ orbitals, as β_1 and β_2. Here $X(Y)$ means the $x(y)$ axis in the original unit cell. Note that in the hole-doped cases, there appears an extra hole Fermi surface around (π, π) called γ.

The phase diagrams of the cuprates and the iron-based superconductors are apparently similar. In both materials, an antiferromagnetic phase is observed in the undoped cases. Doping carriers into the undoped compounds, superconductivity emerges, and the doping dependence of T_c has dome-like feature. Thus superconductivity in the iron-based superconductors considered to be likely being unconventional one. There are experiments reporting that the superconducting gap structure is likely to be s-wave (e.g. Refs. [44–46]). Although the mechanism is still under debate, spin-fluctuation mediated pairing [47, 48] is considered to be a promising candidate since the superconducting phase neighbors the antiferromagnetic phase in many compounds. In the spin-fluctuation mediated pairing mechanism, the Fermi surface nesting between the hole and electron Fermi surfaces plays a key role. Here the term nesting means the degree of the overlap of the Fermi surface translated by a certain vector (nesting vector) with the original Fermi surface. Since the nesting vector roughly corresponds

1.3 Iron-Based Superconductors

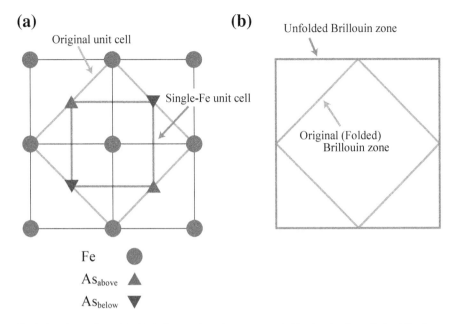

Fig. 1.4 a A top view of the FeAs layer, where As$_{above(below)}$ indicating As above (below) Fe plane. Green and red square indicate the single-Fe and original unit cell. **b** A schematic of the unfolded and original (folded) Brillouin zone (Color figure online)

to a wave vector maximizing spin fluctuation, which give rise to a repulsive pairing interaction, the superconducting gap reversing its sign in the Brillouin zone across the nesting vector connecting the hole and electron Fermi surfaces is likely to be favorable within this mechanism. Based on the analysis for the five-orbital model within the random phase approximation, it has been shown in Ref. [48] that the $s_\pm$-wave gap, where the gap function has the s-wave symmetry but changes its sign in the Brillouin zone as schematically shown in Fig. 1.5b, is obtained within the spin-fluctuation mediated pairing mechanism. As for another mechanism, a pairing interaction exploiting orbital fluctuations towards the structural transition [49, 50], which gives an isotropic gap function without sign changing, is also actively discussed.

As mentioned above, the spin-fluctuation-mediated $s_\pm$-wave pairing mechanism relies on the nesting between the hole and electron Fermi surfaces. However, in some of the iron-based compounds with the optimal T_c higher than 40 K, it is reported that hole Fermi pockets are missing [51–57] as shown in Fig. 1.6, which seems to be detrimental to the spin-fluctuation-mediated pairing since there is no nesting anymore. In this context, the concept of an "incipient band", which is a band lying below (or above), but not far away from, the Fermi level has attracted much attention [41, 55, 58–63]. Namely, exploiting interband scattering processes involving such an incipient band, a pairing interaction can develop in a finite energy region, which

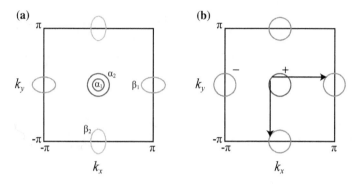

Fig. 1.5 **a** A schematic of the 2D Fermi surface of the five-orbital model. In the panel **b**, a cartoon of $s_\pm$-wave gap and nesting vectors are depicted

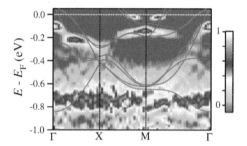

Fig. 1.6 Band structure obtained from (color plot) the ARPES measurement and (red lines) the first principles calculation. In the ARPES band structure, it can be seen that the hole Fermi surfaces around the Γ point (the origin of the Brillouin zone) are missing. Reprinted with permission from Ref. [52], Copyright 2011 by the American Physical Society

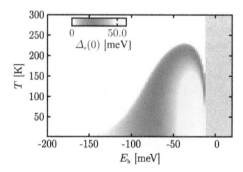

Fig. 1.7 Color plot for the superconducting gap function against the temperature T and the position of the hole band maximum E_h measured from the chemical potential, namely, $E_h < 0$ corresponds to the incipient band case. Gray shaded area indicates the magnetic instability. Reprinted with permission from Ref. [62], Copyright 2016 by the American Physical Society

may lead to an enhancement of superconductivity whose gap function of the electron band has the opposite sign to that of the incipient hole band. This idea was in fact introduced earlier in a different context in Ref. [64], which will be explained in Sect. 2.2.1. In Ref. [62], as shown in Fig. 1.7, it has been semi-phenomenologically shown that spin-fluctuation-mediated superconductivity is enhanced when the hole band can be regarded as incipient in the two-band toy model consisting of the hole- and electron-like parabolic bands.

References

1. Onnes HK (1911) Phys Lab Univ Leiden 119, 120, 133
2. Drozdov A, Eremets M, Troyan I, Ksenofontov V, Shylin S (2015) Nature 525:73
3. Drozdov A et al (2018). arXiv:1808.07039
4. Somayazulu M et al (2018). arXiv:1808.07695
5. Bardeen J, Cooper LN, Schrieffer JR (1957) Phys Rev 108:1175
6. Fournier P (2015) Phys C: Supercond Appl 514:314. Superconducting Materials: Conventional, Unconventional and Undetermined
7. Chu C, Deng L, Lv B (2015) Phys C: Supercond Appl 514:290. Superconducting Materials: Conventional, Unconventional and Undetermined
8. Bednorz JG, Müller KA (1986) Zeitschrift für Physik B Condensed Matter 64:189
9. Wu MK et al (1987) Phys Rev Lett 58:908
10. Schilling A, Cantoni M, Guo J, Ott H (1993) Nature 363:56
11. Momma K, Izumi F (2011) J Appl Crystallogr 44:1272
12. Fukuoka A et al (1997) Phys Rev B 55:6612
13. Mukuda H, Shimizu S, Iyo A, Kitaoka Y (2011) J Phys Soc Jpn 81:011008
14. For some multilayered cases, the situation is somewhat more complicated. For example, an experimental study for a trilayered $Bi_2Sr_2Ca_2Cu_3O_{10}$, Fujii et al (2002) Phys Rev B 66:024507, reported that T_c exhibits an almost constant value against hole doping in the over-doped region, which implies that the doping rate for inequivalent layers is different with each other
15. Yoshida T et al (2003) Phys Rev Lett 91:027001
16. Yoshida T et al (2009) Phys. Rev. Lett. 103:037004
17. Takigawa M et al (1991) Phys Rev B 43:247
18. Sato T et al (2002) Phys Rev Lett 89:067005
19. Vishik IM et al (2010) New J Phys 12:105008
20. Ando Y, Komiya S, Segawa K, Ono S, Kurita Y (2004) Phys Rev Lett 93:267001
21. Naqib S, Uddin MB, Cole J (2011) Phys C: Supercond 471:1598
22. Yamamoto A, Hu W-Z, Tajima S (2000) Phys Rev B 63:024504
23. Tajima S et al (1997) Phys Rev B 55:6051
24. Sato T et al (2000) Phys C: Supercond 341:815
25. Fujimori A et al (2000) Phys C: Supercond 341:2067
26. Dipasupil R, Oda M, Momono N, Ido M (2002) J Phys Soc Jpn 71:1535
27. Torchinsky DH, Mahmood F, Bollinger AT, Božović I, Gedik N (2013) Nat Mater 12:387
28. Ishida K et al (1991) Phys C: Supercond 179:29
29. Imai T, Shimizu T, Yasuoka H, Ueda Y, Kosuge K (1988) J Phys Soc Jpn 57:2280
30. Hammel PC, Takigawa M, Heffner RH, Fisk Z, Ott KC (1989) Phys Rev Lett 63:1992
31. Hardy WN, Bonn DA, Morgan DC, Liang R, Zhang K (1993) Phys Rev Lett 70:3999
32. Shen Z-X et al (1993) Phys Rev Lett 70:1553
33. Ding H et al (1995) Phys Rev Lett 74:2784
34. Alff L et al (1997) Phys Rev B 55:R14757

35. Wollman DA, Van Harlingen DJ, Lee WC, Ginsberg DM, Leggett AJ (1993) Phys Rev Lett 71:2134
36. Tsuei C et al (1996) Phys C: Supercond 263:232
37. Mathai A, Gim Y, Black RC, Amar A, Wellstood FC (1995) Phys Rev Lett 74:4523
38. Iguchi I, Wen Z (1994) Phys Rev B 49:12388
39. Chaudhari P, Lin S-Y (1994) Phys Rev Lett 72:1084
40. Hosono H, Kuroki K (2015) Phys C: Supercond Appl 514:399. Superconducting Materials: Conventional, Unconventional and Undetermined
41. Hirschfeld PJ, Korshunov MM, Mazin II (2011) Rep Progress Phys 74:124508
42. Kamihara Y et al (2006) J Am Chem Soc 128:10012
43. Kamihara Y, Watanabe T, Hirano M, Hosono H (2008) J Am Chem Soc 130:3296
44. Daghero D et al (2009) Phys Rev B 80:060502
45. Hashimoto K et al (2009) Phys Rev Lett 102:017002
46. Nakayama K et al (2009) EPL (Europhys Lett) 85:67002
47. Mazin II, Singh DJ, Johannes MD, Du MH (2008) Phys Rev Lett 101:057003
48. Kuroki K et al (2008) Phys Rev Lett 101:087004
49. Kontani H, Onari S (2010) Phys Rev Lett 104:157001
50. Onari S, Kontani H (2012) Phys Rev Lett 109:137001
51. Guo J et al (2010) Phys Rev B 82:180520
52. Qian T et al (2011) Phys Rev Lett 106:187001
53. Wang Q-Y et al (2012) Chin Phys Lett 29:037402
54. Tan S et al (2013) Nat Mater 12:634
55. Miao H et al (2015) Nat Commun 6:6056
56. Niu XH et al (2015) Phys Rev B 92:060504
57. Total absence of the hole Fermi surface in $K_x Fe_{2-y} Se_2$ is still controversial; in fact, a recent ARPES study, Sunagawa et al (2016) J Phys Soc Jpn 85:073704, observes a "hidden" hole band, which was detected by taking special care of the photon energy and the polarization, likely to be intersecting the Fermi level
58. Wang F et al (2011) EPL (Europhys Lett) 93:57003
59. Bang Y (2014) New J Phys 16:023029
60. Chen X, Maiti S, Linscheid A, Hirschfeld PJ (2015) Phys Rev B 92:224514
61. Bang Y (2016) New J Phys 18:113054
62. Linscheid A, Maiti S, Wang Y, Johnston S, Hirschfeld PJ (2016) Phys Rev Lett 117:077003
63. Mishra V, Scalapino DJ, Maier TA (2016) Sci Rep 6:32078
64. Kuroki K, Higashida T, Arita R (2005) Phys Rev B 72:212509

Chapter 2
Background of the Present Thesis

Abstract In this chapter, we will briefly review the previous studies relevant to the present thesis. We first describe the studies on superconductivity in systems with ladder-type lattice structures. After that, we introduce the studies concerning a mechanism for enhancing superconductivity in systems with coexisting wide and narrow bands. Finally, we describe the FLEX+DMFT method for superconductivity in single-band systems with strong electron correlations, which is a combined method of the non-perturbative dynamical mean-field theory and the perturbative fluctuation exchange approximation.

Keywords Ladder systems · Incipient narrow band · FLEX+DMFT method

2.1 Superconductivity in Ladder Systems

In this section, we briefly review the previous theoretical studies in ladder systems. We also mention the experimental realization of superconductivity in the ladder-type cuprate motivated by them.

2.1.1 Superconductivity in Ladder and Bilayer Models

As mentioned in Chap. 1, models on a square lattice have been extensively studied ever since the discovery of high-T_c cuprates since the Cu atoms form a square lattice. In 1992 and 1993, Dagotto and Rice made an interesting proposal as an alternative way to realize high-T_c superconductivity [1–3]. They considered a ladder-like lattice structure comprising two coupled chains, as depicted in Fig. 2.1a, instead of the square lattice. In Ref. [1], as depicted in Fig. 2.2a, it was shown numerically that the Heisenberg model (and the t-J model) on a two-leg ladder exhibits a spin gap, and the possibility of superconductivity in this spin gap system by doping holes was also discussed by calculating the superconducting correlation as shown in the left panel of Fig. 2.3. Subsequently, in Ref. [2], it has been conjectured that the spin

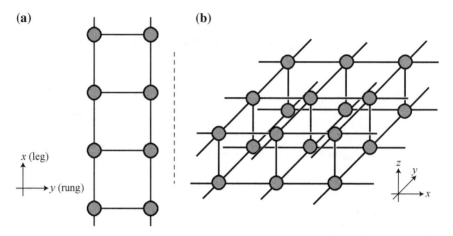

Fig. 2.1 Schematics of lattice structures of **a** a ladder and **b** a bilayer systems

gap opens only in even-leg ladders. Given these studies, in Ref. [3], the conjecture, in which superconductivity occurs by doping carriers into a ladder system with the spin gap, was proposed. Since then, this conjecture has been tested by various analytical methods (e.g. the bosonization technique [4–7] and a mean-field theory using the Gutzwiller renormalization [8]) and numerical methods (e.g. the density matrix renormalization group (DMRG) [9, 10], the projector quantum Monte Carlo method [11], the quantum Monte Carlo [12], and the exact diagonalization (ED) [13]). These studies have shown that the conjecture for superconductivity in the two-leg ladder system is correct, and that the Cooper pairs are formed across the chains.

The bilayer model, whose lattice structure is shown in Fig. 2.1b, can be regarded as a two dimensional analogue of the two-leg ladder model, and was studied in a similar context in Ref. [1]. The calculation results in Ref. [1] for the Heisenberg model (and the t-J model) on a bilayer lattice is similar to those of the two-leg ladder case as shown in Fig. 2.2b and the right panel of Fig. 2.3b. The Hubbard model was studied on a bilayer lattice, which has two identical, bonding and antibonding bands, using the FLEX approximation, which we will describe in Chap. 3 [14, 15]. Also in this system, superconductivity, whose superconducting gap has essentially the same symmetry as that in the two-leg ladder model, is strongly enhanced exploiting finite energy spin fluctuations in the cases of a large bonding-antibonding splitting between two bands, which would be related to an incipient band driven superconductivity as explained in Chap. 4. A dynamical cluster approximation (DCA) study [16] has shown that superconductivity is strongly enhanced in the bilayer Hubbard model which is the same as the model employed in Refs. [14, 15].

On the other hand, since it is known that there is no spin gap in the three-leg ladder model [2, 6, 17], one may expect that superconductivity does not occur in these system according to the conjecture by Dagotto and Rice. However, a weak

2.1 Superconductivity in Ladder Systems

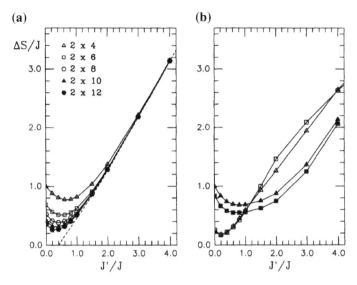

Fig. 2.2 Spin gap as a function of the ratio J'/J for different lattice sizes, where $J^{(l)}$ is the exchange interaction along a leg (rung). Filled (open) symbols indicate the results for the undoped (hole-doped) case. Reprinted with permission from Ref. [1], Copyright 1992 by the American Physical Society

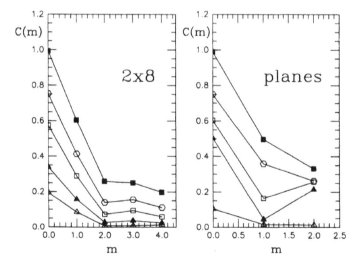

Fig. 2.3 Superconducting correlation as a function of the distance for various strengths of J'. Open triangles, filled triangles, open squares, open hexagons, and filled squares correspond to the results for $J'/J = 0.025, 1, 4, 10$, and 100, respectively. Reprinted with permission from Ref. [1], Copyright 1992 by the American Physical Society

coupling study with the bosonization at the renormalization-group fixed point [18, 19] revealed that superconducting pairing across the central and edge chains becomes dominant. In this system, actually, two of three spin excitation modes are gapfull while the other is gapless. Because of this, superconducting pairing can be dominant using the gapfull modes at least in the weak-coupling region.

2.1.2 Experiment on Two-Leg Ladder Compounds

As a typical series of compounds with ladder-type lattice structures, n-chain cuprates $Sr_{n-1}Cu_nO_{2n-1}$, e.g. a two-leg ladder compound $SrCu_2O_3$ and a three-leg $Sr_2Cu_3O_5$ [20], were synthesized in the past. In these materials, inter-ladder couplings are considered to be quite weak since ladders are connected by Cu-O-Cu bonds with a 90° of bond angle. As for magnetic properties, susceptibility and NMR measurements have shown that a two-leg ladder compound $SrCu_2O_3$ is a spin gap system [21, 22] while a three-leg ladder compound $Sr_2Cu_3O_5$ does not have the spin gap [22], which is consistent with theories of two- and three-leg ladder. However, since these materials are known to be unable to dope carriers, it seems difficult to have superconductivity in these materials.

Another typical ladder cuprate is $Sr_{14-x}Ca_xCu_{24}O_{41}$ having the crystal structure where Cu-O chains and Cu-O ladders are alternately stacked [23], which is also shown to be a spin gap system for $x = 0$ by a NMR measurement [24]. By doping Ca and applying pressure, holes are doped into ladders from chains [25]. Because of this, the system becomes a metal and superconductivity has been observed [23] with $T_c \sim 10$ K. This apparently seems to be an experimental realization of the theoretically conjectured superconductivity in a ladder system. However, since it is experimentally known that the two-dimensional character increases by applying pressure, superconductivity in this compound can be regarded as that of a two-dimensional system and hence it is possible that it is not completely the same as the theoretically conjectured one. To treat this problem, perturbative analysis based on the fluctuation exchange (FLEX) approximation for an effective two-dimensional model of this compound has been performed [26], which gives a sign-reversing anisotropic superconducting gap arising from spin fluctuations.

In any case, the superconducting T_c of the ladder-type cuprates so far observed is not so high as those of the layered cuprates. In the context of the proposal by Dagotto and Rice, T_c becomes high when the exchange J' in the rung direction becomes large (in Fig. 2.3, larger superconducting correlation corresponds to higher T_c). This corresponds to the case when the electron hopping amplitude in the rung direction is larger than that in the leg direction. In the actual ladder cuprates, however, it is known that the hopping, and hence the exchange, in the rung direction is slightly smaller than that in the leg direction. Hence, a realization of T_c as high as or even exceeding those of the layered cuprates may seem difficult in the actual cuprate ladder compounds. This can also be seen from the FLEX calculation of the Hubbard model on a two-leg ladder lattice in Ref. [26].

2.2 Superconductivity in Systems with Coexisting Wide and Narrow Bands

In the previous section, we have reviewed the previous studies for ladder systems. Although there were some expectations for high-T_c in ladder-type cuprates, the experimentally observed T_c so far is not so high as those of the layered (square lattice) cuprates. In 2005, a subsequent proposal was made, which suggested that a much higher T_c may take place in two-leg cuprate ladder compounds in a certain condition [27]. In this section, we shall describe the idea of the proposal and related studies.

2.2.1 An Idea for Enhancing T_c

While the large energy scale t of electrons naively favors high-T_c in electron mechanisms of superconductivity, we usually end up with T_c as low as $\sim t/100$. For the cuprates, which are typical unconventional superconductors, since the energy scale is $t \sim \mathcal{O}(1)$ eV, $T_c \sim t/100$ (~ 100 K) is still high. On the other hand, T_c of the conventional phonon-mediated superconductors is about several percent of the energy scale of phonons. Given this, an important question is how we can design electron-mechanism superconductivity where the "low" values of T_c could be enhanced.

An ideal situation for enhancing T_c in the unconventional superconductors is to simultaneously have a strong pairing interaction and a light renormalized electron mass, but usually the two are not compatible with each other. Namely, a strong electron correlation can mediate strong pairing interactions, but usually makes the electron mass heavy due to the strong quasi-particle renormalization. For instance, let us consider a spin-fluctuation mediated pairing in the single-band Hubbard model on a square lattice, which is a simplest model of the high-T_c cuprates. In this case, we have indeed a strong pairing interaction simultaneously accompanied by the strong quasiparticle renormalization. The former works in favor of superconductivity, whereas the latter works against it. As a way to overcome this problem, we can consider multi-band systems with coexisting wide and narrow (or even flat) bands as proposed in the previous studies e.g. Refs. [27–29].

Reference [27] proposed a possible enhancement of superconductivity exploiting coexisting wide- and narrow bands. It requires a system with narrow- and wide-bands, not due to multiple orbitals like in the heavy fermion compounds, but due to multiple atomic sites in a unit cell such as a ladder with diagonal hopping t'. Let us consider the situation in which the Fermi level lies close to, but not right within, the narrow band, as schematically shown in Fig. 2.4. The electrons in the wide band can form Cooper pairs, which are not so strongly renormalized, with a pairing interaction strongly mediated by the large number of interband scattering channels, where the narrow band acts as intermediate states. This is favorable for high-T_c superconductivity. Note that the narrow band in this case can be regarded

Fig. 2.4 A schematic of coexisting wide and incipient narrow bands

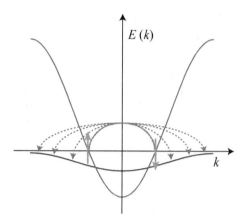

as a prototype of an "incipient band", a terminology introduced in the context of iron-based superconductors.

2.2.2 Previous Studies on Systems with Coexisting Wide and Narrow Bands

In Ref. [27], the Hubbard model on a two-leg ladder lattice was considered as a way to realize the above mentioned system with coexisting wide and incipient narrow bands. Namely, the band structure of the tight-binding model on a two-leg ladder exhibits bonding and anti-bonding bands, and when electrons can hop only to nearest neighbors, the two bands have the same band width, while one of the bands become wide and the other narrow when the next-nearest-neighbor hoppings are introduced. Applying the FLEX approximation to this model, it was predicted that $T_c \sim 200$ K takes place in the two-leg ladder cuprate, provided that 30% of electron doping is achieved, so that the Fermi level is placed just above the narrow band edge and hence the incipient band situation is realized (Fig. 2.5). Although the absolute value of $T_c = 200$ K itself may not be reliable quantitatively, it is certain that the calculated T_c is much higher than the one obtained for the Hubbard model on a square lattice using the same FLEX approximation (compare Fig. 2.5 and the left panels of Fig. 2.9). The obtained superconductivity has the s-wave gap symmetry having opposite sign between two bands but without sign changing within the band, and the dominant component is that corresponding the pairs across a rung of a ladder. A problem, however, is that doping carriers into the ladder cuprates is known to be difficult [30]; especially electron doping with such a large amount may be rather unrealistic.

In Refs. [29, 31], systematic calculations for systems with coexisting wide and narrow bands within the FLEX approximation have been performed. It has been found that a pairing mechanism exploiting an incipient narrow band is applicable to a wide variety of models in quasi-one, two, and three dimensions.

2.2 Superconductivity in Systems with Coexisting Wide and Narrow Bands

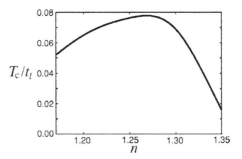

Fig. 2.5 Carrier concentration dependence of T_c for the Hubbard model on a two leg ladder with diagonal hopping t'. t_l denotes the hopping integral along a leg of a ladder, which is typically ~0.4 eV in the ladder cuprate, and n denotes the number of electrons per site. Reprinted with permission from Ref. [27], Copyright 2005 by the American Physical Society

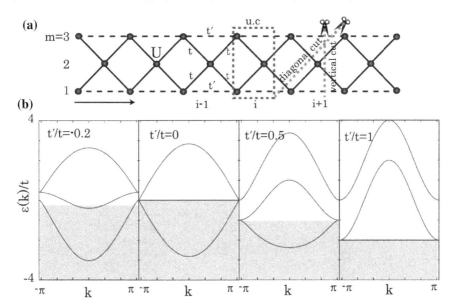

Fig. 2.6 a Lattice and **b** non-interacting band structures of the diamond chain model. t' denotes the interapex hopping integral, and for $t' = O(1)$, the middle (bottom) band becomes perfectly flat. Reprinted with permission from Ref. [28], Copyright 2016 by the American Physical Society

However, since the above-mentioned studies adopt the FLEX approximation, which is based on the perturbation theory with respect to the interaction term, it is still unclear how the strong correlation effects affect this scenario. Both local and non-local (e.g. magnetic/charge fluctuations) correlations should be incorporated for treating this kind of pairing mechanisms since we have to consider a pairing inter-

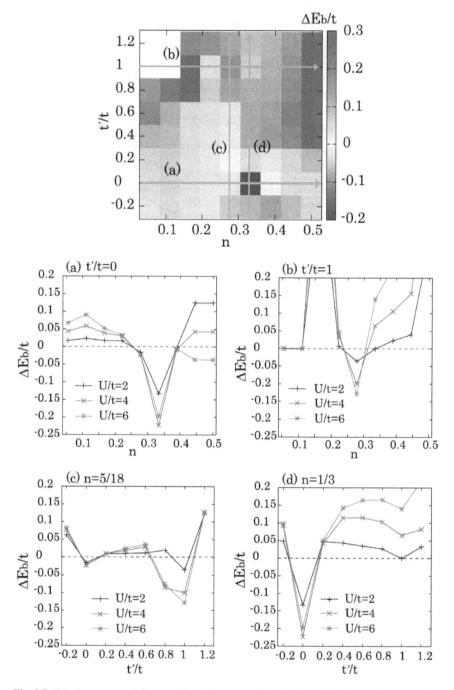

Fig. 2.7 Binding energy of Cooper pairs. The top panel depicts the color plot against n and $t^p r$, where t' is the hopping shown in Fig. 2.6 and n is the number of electrons per site per spin and (a–d) some cuts of that along a constant n or $t^p r$. Reprinted with permission from Ref. [28], Copyright 2016 by the American Physical Society

2.2 Superconductivity in Systems with Coexisting Wide and Narrow Bands

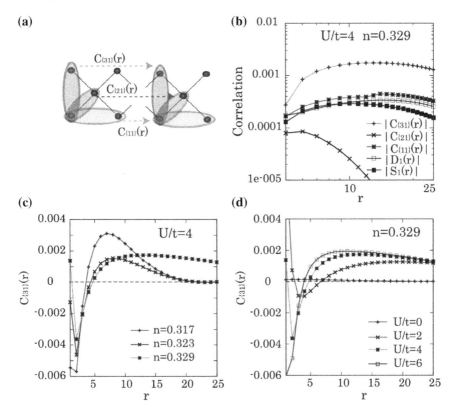

Fig. 2.8 Superconducting correlation function against real space distance for various pair structures shown in the panel (**a**). D and S are the density and spin correlation functions, respectively. Reprinted with permission from Ref. [28], Copyright 2016 by the American Physical Society

action mediated by non-local (magnetic) correlations in the presence of the strong local correlation.

There are some studies for these kind of systems based on non-perturbative methods. In Ref. [28], superconductivity in the Hubbard model on a diamond chain, depicted in Fig. 2.6, was studied using ED and DMRG [28]. In this model, one of the bands become perfectly flat under certain conditions, as shown in Fig. 2.6b. ED gives a significant binding energy of Cooper pairs when the Fermi level lies in the vicinity of the flat band as shown in Fig. 2.7. DMRG calculation was also performed, and long-ranged the Cooper pair correlation was obtained at the same condition for enhancement of the binding energy of Cooper pairs (Fig. 2.8). This implies an enhancement of superconductivity in a system with wide and flat band if the Fermi level lies in the vicinity of the flat band. Although these methods are numerically exact, the study has been performed only for a quite specific situation, i.e. a purely one-dimensional lattice, at zero temperature, and with a perfectly flat band. Therefore

Fig. 2.9 Band filling dependence of T_c obtained with (left panel) the FLEX approximation and (right) FLEX+DMFT method. Filling is defined as the number of electrons per unit cell and filling= 1 corresponds to the half-filled case. The panel (**a**) is result for the single-band Hubbard model on a square lattice only with the nearest neighbor hopping, and the panel (**b**) is for the single-band Hubbard model on a square lattice with the next nearest neighbor and the third neighbor hoppings. Reprinted with permission from Ref. [33], Copyright 2015 by the American Physical Society

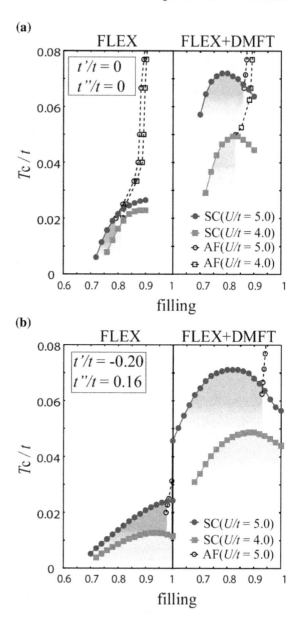

a systematic analysis on the role of strong correlations due to the strong interband scattering is highly desired.

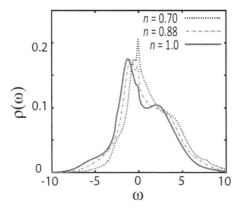

Fig. 2.10 Spectral functions of the single-band Hubbard model on a square lattice with the next nearest neighbor and the third neighbor hoppings for some carrier concentrations n obtained with the FLEX+DMFT method. At half-filling ($n = 1$), there can be seen a dip structure at low energy. Reprinted with permission from Ref. [33], Copyright 2015 by the American Physical Society

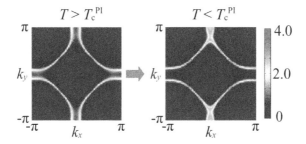

Fig. 2.11 Fermi surfaces of the single-band Hubbard model (left) above and (right) below the Pomeranchuk instability temperature T_c^{PI} obtained with the FLEX+DMFT method. Reprinted with permission from Ref. [34], Copyright 2017 by the American Physical Society

2.3 FLEX+DMFT Method for Single-Band Systems

In order to perform a systematic analysis on the role of strong correlations due to the strong interband scattering, a method treating both local and non-local correlations self-consistently with a moderate computational cost is necessary. To this end, the FLEX+DMFT approach [32–34], namely a method combining the FLEX approximation, which treats non-local correlations perturbatively, and the dynamical mean-field theory (DMFT), which describe local correlations *non-perturbatively*, would be promising. Its computational cost is comparable with that of DMFT, which is cheaper than that of other methods incorporating both local and non-local correlations, since the FLEX approximation is computationally much cheaper than DMFT. Since our aim is superconductivity in multi-band systems, we have to generalize this to multi-band systems, which will be done in Chap. 3. In this section, we briefly review the previous studies for superconductivity and correlation effects in the single-band Hubbard model on a square lattice [33, 34].

In Ref. [33], the doping dependence of T_c in the Hubbard model on a square lattice, which is a simplest model of the cuprates, was studied. The obtained doping dependence exhibits a dome like feature, which can not be reproduced by the FLEX approximation as shown in Fig. 2.9, and it has been shown that its origin is the doping dependence of the electron correlation effects and the pairing interaction arising from corrections to the DMFT part. In addition, a dip structure in the spectral function, i.e. the density of states involving correlation effects, has been found, namely the method can capture the pseudogap behavior as shown in Fig. 2.10, which is considered as a typical manifestation of the strong correlation in the cuprates.

In Ref. [34], it was shown that this method can describe the Pomeranchuk instability as shown in Fig. 2.11, which is an electron-correlation-induced Fermi surface deformation from the tetragonal to orthorhombic symmetry. The Pomeranchuk instability was proposed in Ref. [35] using the functional renormalization group method for the Hubbard model on a square lattice and in Ref. [36] using the slave-boson mean-field theory for the t-J model on a square lattice.

Since the FLEX+DMFT method can capture some intriguing properties of the single-band Hubbard model as mentioned here, we can expect that this method would be a promising approach also for multi-band systems.

References

1. Dagotto E, Riera J, Scalapino D (1992) Phys Rev B 45:5744
2. Rice TM, Gopalan S, Sigrist M (1993) EPL (Europhys Lett) 23:445
3. Dagotto E, Rice TM (1996) Science 271:618
4. Khveshchenko DV, Rice TM (1994) Phys Rev B 50:252
5. Khveshchenko DV (1994) Phys Rev B 50:380
6. Schulz HJ (1996) Phys Rev B 53:R2959
7. Balents L, Fisher MPA (1996) Phys Rev B 53:12133
8. Sigrist M, Rice TM, Zhang FC (1994) Phys Rev B 49:12058
9. Noack RM, White SR, Scalapino DJ (1994) Phys Rev Lett 73:882
10. Noack RM, White SR, Scalapino DJ (1995) EPL (Europhys Lett) 30:163
11. Asai Y (1995) Phys Rev B 52:10390
12. Kuroki K, Kimura T, Aoki H (1996) Phys Rev B 54:R15641
13. Tsunetsugu H, Troyer M, Rice TM (1994) Phys Rev B 49:16078
14. Kuroki K, Kimura T, Arita R (2002) Phys Rev B 66:184508
15. Nakata M, Ogura D, Usui H, Kuroki K (2017) Phys Rev B 95:214509
16. Maier TA, Scalapino DJ (2011) Phys Rev B 84:180513
17. Arrigoni E (1996) Phys Lett A 215:91
18. Kimura T, Kuroki K, Aoki H (1996) Phys Rev B 54:R9608
19. Lin H-H, Balents L, Fisher MPA (1997) Phys Rev B 56:6569
20. Hiroi Z, Azuma M, Takano M, Bando Y (1991) J Solid State Chem 95:230
21. Azuma M, Hiroi Z, Takano M, Ishida K, Kitaoka Y (1994) Phys Rev Lett 73:3463
22. Ishida K et al (1996) Phys Rev B 53:2827
23. Uehara M et al (1996) J Phys Soc Jpn 65:2764

References

24. Tsuji S, Kumagai K-I, Kato M, Koike Y (1996) J Phys Soc Jpn 65:3474
25. Matsuda M, Katsumata K (1996) Phys Rev B 53:12201
26. Kontani H, Ueda K (1998) Phys Rev Lett 80:5619
27. Kuroki K, Higashida T, Arita R (2005) Phys Rev B 72:212509
28. Kobayashi K, Okumura M, Yamada S, Machida M, Aoki H (2016) Phys Rev B 94:214501
29. Matsumoto K, Ogura D, Kuroki K (2018) Phys Rev B 97:014516
30. Kojima K, Motoyama N, Eisaki H, Uchida S (2001) J Electron Spectrosc Relat Phenom 237:117–118
31. K. Matsumoto, D. Ogura, and K. Kuroki (unpublished)
32. Gukelberger J, Huang L, Werner P (2015) Phys Rev B 91:235114
33. Kitatani M, Tsuji N, Aoki H (2015) Phys Rev B 92:085104
34. Kitatani M, Tsuji N, Aoki H (2017) Phys Rev B 95:075109
35. Halboth CJ, Metzner W (2000) Phys Rev Lett 85:5162
36. Yamase H, Kohno H (2000) J Phys Soc Jpn 69:2151

Chapter 3
Method

Abstract In this chapter, we will describe theoretical and computational methods used in the present study. Firstly, we introduce the tight-binding model, which is a convenient starting point of a many-body theory for treating electron correlation effects. Then, we introduce a many-body theory for treating electron correlation effects based on the Green's function method and some approximations which are relevant to this study. We also introduce the first-principles method based on the density-functional theory, which is a powerful tool for calculations of actual materials. Finally, the maximally localized Wannier function method for constructing effective tight-binding model from the first-principles result is also described.

Keywords Density functional theory · Fluctuation exchange approximation · Dynamical mean-field theory

3.1 Tight-Binding Model

There are two pictures to describe the electronic structure of solids. One is that the band structures arise from the modification upon free electrons represented by plane waves due to the periodic potential of a crystal. The other is that the band structures originate from the processes where electrons hop from an atomic site to neighboring sites due to the overlap of wave functions for isolated atoms located at each atomic site. Typically, the former is a weak coupling picture, whereas the latter is a strong coupling one. Since we focus on superconductivity arising from the electron correlation, here we explain the latter.

3.1.1 Wannier Functions

The Wannier functions form a complete set of orthogonal functions and are localized around atomic sites, and hence are convenient basis functions for expressing the electronic structure. The Wannier functions $w_{Rn}(r)$ are defined by the inverse Fourier series of the Bloch functions $\psi_{nk}(r)$:

$$w_{Rn}(r) = \langle r | w_R \rangle, \tag{3.1}$$

$$|w_{Rn}\rangle = \frac{V}{(2\pi)^2} \sum_k^{\text{B.Z.}} \sum_m e^{-ik \cdot r} U_{mn}^{(k)} |\psi_{mk}\rangle, \tag{3.2}$$

$$\psi_{nk}(r) = \langle r | \psi_{nk} \rangle = u_{nk}(r) e^{ik \cdot r}. \tag{3.3}$$

Here U^k is a unitary matrix transforming the basis (e.g band ↔ orbital), and has many different choices. In the isolated atomic limit, the Wannier functions approach to the atomic orbitals.

3.1.2 Tight-Binding Model

The kinetic part of the Hamiltonian with the orbital degrees of freedom $\mathcal{H}_{\text{kin}}$ is expressed as,

$$\mathcal{H}_{\text{kin}} = \sum_\sigma \sum_{\mu\nu} \int dr \, \psi_\sigma^{\mu\dagger}(r) T(r) \psi_\sigma^\nu(r), \tag{3.4}$$

where, $T(r)$ is the one-body kinetic term, $\mu(\nu)$ denotes orbitals. $\psi_\sigma^{\mu\dagger}(r)$, $\psi_{\mu\sigma}(r)$ are creation and annihilation field operators of electrons with spin σ on the μth orbital, respectively. Field operators ψ can be expanded using the Wannier functions ϕ^μ:

$$\psi_\sigma^{\mu\dagger}(r) = \sum_i \phi^{\mu*}(r - R_i) c_{i\sigma}^{\mu\dagger} \tag{3.5}$$

$$\psi_\sigma^\mu(r) = \sum_i \phi^\mu(r - R_i) c_{i\sigma}^\mu. \tag{3.6}$$

$c_{i\sigma}^{\mu\dagger}$ ($c_{i\sigma}^\mu$) creates (annihilates) an electron with spin σ on the μth orbital at the ith site. Therefore we can rewrite $\mathcal{H}_{\text{kin}}$ as,

$$\mathcal{H}_{\text{kin}} = \sum_{ij} \sum_\sigma \sum_{\mu\nu} t_{ij}^{\mu\nu} c_{i\sigma}^{\mu\dagger} c_{j\sigma}^\nu, \tag{3.7}$$

with the hopping integrals,

$$t_{ij}^{\mu\nu} = \int dr \, \phi^{\mu*}(r - R_i) T(r) \phi^\nu(r - R_j), \tag{3.8}$$

which means the amplitude of a hopping process from the νth orbital at the jth site to the μth orbital at the ith site. This model is called the tight-binding model.

3.1 Tight-Binding Model

In the k-space, the Hamiltonian becomes,

$$\mathcal{H} = \sum_{k,\sigma} \sum_{\mu\nu} \mathcal{H}_{\mu\nu}(k) c_{k\sigma}^{\mu\dagger} c_{k\sigma}^{\nu}, \tag{3.9}$$

where,

$$c_{k\sigma}^{\mu} = \frac{1}{\sqrt{N}} \sum_{i} \exp(-i k \cdot r_i) c_{i\sigma}^{\mu}. \tag{3.10}$$

To note that here N denotes the number of atomic sites. $\mathcal{H}_{\mu\nu}(k)$ is the Fourier series of $t_{ij}^{\mu\nu}$, and one can obtain the band structure by diagonalizing this.

In addition to this, for strongly correlated electron systems, one has to consider the electron–electron interaction with the Hamiltonian,

$$\mathcal{H}_{\text{int}} = \frac{1}{2} \sum_{\alpha\beta\gamma\delta} \sum_{i,j} \sum_{\sigma\sigma'} V_{\alpha\beta\gamma\delta}(r_i - r_j) c_{i\sigma}^{\alpha\dagger} c_{j\sigma'}^{\beta\dagger} c_{j\sigma'}^{\delta} c_{i\sigma}^{\gamma}. \tag{3.11}$$

It is necessary to simplify this Hamiltonian in order to assimilate the effects of the interaction. Considering that the Coulomb interaction in the many-electron systems is strongly short-ranged due to the screening effect, we can approximate the interaction so as to act only when two electrons occupy the same atomic site:

$$\mathcal{H}_{\text{int}} = \sum_{i} \left(\sum_{\mu} U_{\mu} n_{i\mu\uparrow} n_{i\mu\downarrow} + \sum_{\mu > \nu} \sum_{\sigma\sigma'} U'_{\mu\nu} n_{i\mu\sigma} n_{i\mu\sigma'} \right.$$
$$\left. - \sum_{\mu \neq \nu} J_{\mu\nu} S_{i\mu} \cdot S_{i\nu} + \sum_{\mu \neq \nu} J'_{\mu\nu} c_{i\uparrow}^{\mu\dagger} c_{i\downarrow}^{\mu\dagger} c_{i\downarrow}^{\nu} c_{i\uparrow}^{\nu} \right),$$

where $n_{i\mu\sigma}$ is the number operator for an electron with spin σ on the μth orbital at the ith site and $S_{i\mu}$ is the spin operator for an electron on the μth orbital at the ith site. $U^{(\prime)}$ is the intra(inter)orbital interaction and J denotes the Hund's coupling as well as J' denotes the pair hopping. The model considering this on top of the tight-binding model is called the multi-orbital Hubbard model, which is widely used in the study for strongly correlated electron systems.

The single band version of the Hubbard model often studied as a model for the high-T_c cuprates, as mentioned in Chap. 1, is given as

$$\mathcal{H} = \sum_{ij} \sum_{\sigma} t_{ij} c_{i\sigma}^{\dagger} c_{j\sigma} + \sum_{i} U n_{i\uparrow} n_{i\downarrow}. \tag{3.12}$$

3.2 Green's Function Method for Many-Body Problem

It is most important in this study to treat the many-body effects in the constructed model. In this section, we describe the general formalism using the Green's function and the microscopic theories of superconductivity, the BCS theory and the Eliashberg theory, referencing Refs. [1–4]. For simplicity, we treat the single-orbital case in this section. However, generalization for systems with internal degrees of freedom (e.g. orbital or sublattice) can be done in a straight forward manner by interpreting the Green's function as a matrix.

3.2.1 Green's Function

Let us define the Heisenberg representation of an operator A:

$$\bar{A}(u) = \exp(u\mathcal{H}) A \exp(-u\mathcal{H}), \tag{3.13}$$

where u corresponds to inverse temperature $\beta = 1/k_B T$ and is called imaginary time from a formal analogy with the Heisenberg representation in real time.

Using this representation, let us define the (thermal) Green's function:

$$G_{AB}(u, u') = -\langle T_u \bar{A}(u) \bar{B}(u') \rangle. \tag{3.14}$$

Here $\langle A \rangle$ is the statistical average of A, and $T_u \bar{A}(u_1) \bar{B}(u_2) \ldots$ is called the time-ordered product, which sorts operators in a descending sequence in imaginary time and multiply a factor of -1 for each interchange of fermion operators. One can easily show the following properties:

$$G_{AB}(u, u') = G_{AB}(u - u'), \tag{3.15}$$

$$G_{AB}(u + \beta) = \pm G_{AB}(u), \tag{3.16}$$

where the double sign corresponds to bosons (upper) and fermions (lower). Namely, $G_{AB}(u)$ is periodic/anti-periodic with the period of β and hence can be expressed by the Fourier series.

$$G_{AB}(u) = \frac{1}{\beta} \sum_{\omega_n} G(i\omega_n) \exp(-i\omega_n u), \tag{3.17}$$

$$G_{AB}(i\omega_n) = \frac{1}{2} \int_{-\beta}^{\beta} du\, G_{AB}(u) \exp(i\omega_n u)$$

$$= \int_{0}^{\beta} du\, G_{AB}(u) \exp(i\omega_n u), \tag{3.18}$$

3.2 Green's Function Method for Many-Body Problem

where ω_n is called Matsubara frequency satisfying

$$\omega_n = \begin{cases} 2n\pi k_B T & \text{for bosons} \\ (2n+1)\pi k_B T & \text{for fermions,} \end{cases} \quad (3.19)$$

for integer n.

Specifically for $A = c_{k\sigma}$, $B = c_{k\sigma}^{\dagger}$, $G_{AB}(u)$ is a one-body Green's function which is a key quantity in treating electron correlation effects. Hereafter, we simply call it the Green's function and denote it as $G_{\sigma}(k, u)$.

Green's function for the non-interacting case $G^{(0)}$ can be written as

$$G_{\sigma}^{(0)}(k) = -\int_0^{\beta} du \exp\left((i\omega_n - \xi(k))u\right)(1 - f_k),$$
$$= \frac{1}{i\omega_n - \xi(k)}, \quad (3.20)$$

here $\xi(k)$ is the energy dispersion measured from the chemical potential μ and $k \equiv (k, i\omega_n)$ is a shorthand for the wavenumber and the Matsubara frequency.

3.2.2 Lehmann Representation and Retarded Green's Function

Here we rewrite the Green's function $G_{AB}(u)$ in terms of the eigenstates $\{|\alpha\rangle\}$. Since $|\alpha\rangle$ satisfies

$$\mathcal{H}|\alpha\rangle = E|\alpha\rangle, \quad (3.21)$$

$G(i\omega_n)$ can be transformed as

$$G_{AB}(i\omega_n) = -\int_0^{\beta} du \, \text{Tr}\left\{\exp\left(\beta(\Omega - H)\right) \bar{A}(u)\bar{B}(0)\exp(i\omega_n u)\right\}$$
$$= -\int_0^{\beta} du \sum_{\alpha',\alpha''} \exp\left(\beta(\Omega - E')\right)\exp\left(u(E' - E'' + i\omega_n)\right)\langle\alpha'|A|\alpha''\rangle\langle\alpha''|B|\alpha'\rangle$$
$$= -\sum_{\alpha',\alpha''} \exp\left(\beta(\Omega - E')\right)\frac{\pm\exp\left(\beta(E' - E'')\right) - 1}{i\omega_n + E' - E''}\langle\alpha'|A|\alpha''\rangle\langle\alpha''|B|\alpha'\rangle . qquad$$
$$(3.22)$$

Here $\Omega = -k_B T \ln \Xi$ is the thermodynamic potential with the grand partition function Ξ. Using the spectral function $A_{AB}(\omega)$ defined by

$$A_{AB}(\omega) = 2\pi \sum_{\alpha',\alpha''} \exp\left(\beta(\Omega - E')\right)(\exp(\beta\omega) \mp 1)$$
$$\times \langle\alpha'|B|\alpha''\rangle\langle\alpha''|A|\alpha'\rangle \delta\left(\omega - (E' - E'')\right), \quad (3.23)$$

we obtain the following integral representation called the Lehmann representation:

$$G_{AB}(i\omega_n) = \frac{1}{2\pi} \int_{-\infty}^{\infty} d\omega \frac{A_{AB}(\omega)}{i\omega_n - \omega} \quad (3.24)$$

To clarify the relationship between the Green's function for Matsubara frequency and that for real frequency, we introduce the retarded Green's function using the Heisenberg representation in real time:

$$G_{AB}^r(t) = -i\theta(t)\left\langle \left[\bar{A}(t), \bar{B}(0)\right]_{\mp} \right\rangle \quad (3.25)$$

Again, this can be written in terms of the eigenstates:

$$G_{AB}^r(t) = -i\theta(t) \sum_{\alpha',\alpha''} \exp\left(\beta(\Omega - E')\right) \exp\left(-i(E' - E'')t\right)$$
$$\times \left[\exp\left(\beta(E' - E'')\right) \mp 1\right] \langle\alpha'| A |\alpha''\rangle \langle\alpha''| B |\alpha'\rangle . \quad (3.26)$$

Since $G_{AB}^r(t) = 0$ for $t < 0$,

$$G_{AB}^r(\omega) = \int_0^{\infty} dt \exp(i(\omega + i\delta')t) G_{AB}^r(t). \quad (3.27)$$

Here we have introduced an infinitesimal positive number δ to assure the convergence of the integral.

$$G_{AB}^r(\omega) = \int_{-\infty}^{\infty} d\omega' \frac{A_{AB}(\omega')}{\omega + i\delta - \omega'}. \quad (3.28)$$

Therefore we obtain a correspondence,

$$G_{AB}^r(\omega) = G_{AB}(\omega + i\delta). \quad (3.29)$$

Namely, we can obtain information on real frequency by performing "analytic continuation" from the Matsubara representation to the real frequency representation using $G_{AB}(i\omega_n)$.

In the present study, we employ the Padé approximation and the maximum entropy method to obtain the spectral function by the analytic continuation. We describe basics for these methods in Appendix A.

3.2.3 Thermodynamic Potential

For later discussion, let us decompose the Hamiltonian $\mathcal{H}$ as

3.2 Green's Function Method for Many-Body Problem

$$\mathcal{H} = \mathcal{H}_0 + \mathcal{H}_1, \tag{3.30}$$

where, $\mathcal{H}_0$ is the unperturbed part and usually the kinetic term. Here we take the two-body interaction as $\mathcal{H}_1$, which is treated as perturbation.

Let us introduce the interaction representation of an operator $A(u)$ by replacing $\mathcal{H}$ by $\mathcal{H}_0$. $\bar{A}(u)$ and $A(u)$ have the relation

$$\bar{A}(u) = U(u)^{-1} A(u) U(u), \tag{3.31}$$

where $U(u)$

$$U(u) = \exp(u\mathcal{H}_0)\exp(-u\mathcal{H}). \tag{3.32}$$

Let us define $S(u, u')$ by

$$S(u, u') = U(u)U(u')^{-1}. \tag{3.33}$$

Taking u derivative, one can obtain

$$-\frac{\partial S(u, u')}{\partial u} = \mathcal{H}_1(u) S(u, u'). \tag{3.34}$$

With this, we can obtain an integral equation

$$S(u, u') = \sum_{n=0}^{\infty} \frac{(-1)^n}{n!} \int_{u'}^{u} du_1 \ldots \int_{u'}^{u} du_n T_u[\mathcal{H}_1(u_1) \ldots \mathcal{H}_n(u_1)]. \tag{3.35}$$

Or formally,

$$S(u, u') = T_u \exp\left(-\int_{u'}^{u} du_1 \mathcal{H}_1(u_1)\right). \tag{3.36}$$

Therefore the perturbative expansion of the thermodynamic potential can be expressed as

$$\Omega = \Omega_0 - k_B T \ln \left\langle T_u \exp\left(-\int_0^{\beta} du \mathcal{H}_1(u)\right)\right\rangle_0. \tag{3.37}$$

Here Ω_0 means the non-interacting thermodynamic potential and $\langle T_u \ldots \rangle_0$ the statistical average using $\mathcal{H}_0$.

It can be seen in Eq. (3.37) that a calculation of the thermodynamic potential necessitate to evaluate the statistical average of the product of a lot of creation/annihilation operators. Upon this, the Bloch–de Dominicis theorem is very useful, which enables us to transform the product into a tractable form. The Bloch–de Dominicis theorem can be expressed as

$$\langle C_1 C_2 \ldots C_{2n} \rangle_0 = \sum_{\{i_1 i_2 \ldots i_{2n}\}} \text{sgn}\mathcal{P} \langle C_{i_1} C_{i_2} \rangle_0 \langle C_{i_3} C_{i_4} \rangle_0 \ldots \langle C_{i_{2n-1}} C_{i_{2n}} \rangle_0. \qquad (3.38)$$

Here C corresponds to $c_{k\sigma}(u)$ or $c_{k\sigma}^{\dagger}(u)$. The summation is taken for the permutations under the constraint $i_1 < i_2, i_3 < i_4, \ldots, i_{2n-1} < i_{2n}$ and $i_1 < i_2 < \cdots < i_{2n-1}$. $\text{sgn}\mathcal{P}$ is the sign of the permutation $\mathcal{P}$ which rearranges the product in the left hand side to agree with the order on the right hand side. Namely, the statistical average of the product of a lot of creation/annihilation operators can be expressed in terms of the product of the statistical average of two creation/annihilation operators.

Using Eq. (3.38), each term in Eq. (3.37) can be expressed by a product of $\langle c^{\dagger} c \rangle_0$. But, as we will see here, it is not necessary to consider all the combinations of the Bloch–de Dominicis decomposition. We call a term in which imaginary time arguments in two operators differ from each other a connected term, whereas we call a term in which imaginary time arguments of two operators coincides with each other, namely an imaginary-time-independent term a disconnected term. Note that the first order term $S^{(1)}$ of $\langle S(\beta, 0) \rangle_0$ is regarded as connected in convenience. Every disconnected term can be written in terms of a product of lower-order terms than it, e.g. $S_c^{(n_1)^{m_1}} \ldots S_c^{(n_j)^{m_j}}$. Furthermore, counting the combination, one can see that its coefficient is $1/m_1! \cdots m_j!$. Adding all the contribution up, Eq. (3.37) can be reduced to

$$\Omega = \Omega_0 - k_B T \left[\left\langle T_u \exp\left(-\int_0^{\beta} du \mathcal{H}_1(u)\right) \right\rangle_{0,c} - 1 \right]. \qquad (3.39)$$

The subscript c means that we consider only connected terms. Namely, in the calculation of Ω, it is sufficient to collect only connected terms. This is called the linked cluster theorem.

3.2.4 Feynman Diagram

It is necessary to deal with a lot of terms in the perturbative expansion of Ω due to the Bloch-de Dominicis decomposition. Here we introduce the Feynman diagrams, which enable us to collect perturbation terms systematically by diagrammatically denoting them.

For example, the first order contribution $\Omega^{(1)}$ of Ω is composed of two terms $\Omega^{(1)} = \Omega^{(1a)} + \Omega^{(1b)}$:

$$\Omega^{(1a)} = \frac{(k_B T)^2}{2} \sum_{kk'} \sum_{\epsilon_n \epsilon_{n'}} \sum_{\sigma \sigma'} V(\mathbf{0}) G_{\sigma}^{(0)}(k) \exp(i\epsilon_n \delta) G_{\sigma'}^{(0)}(k') \exp(i\epsilon_{n'} \delta), \qquad (3.40)$$

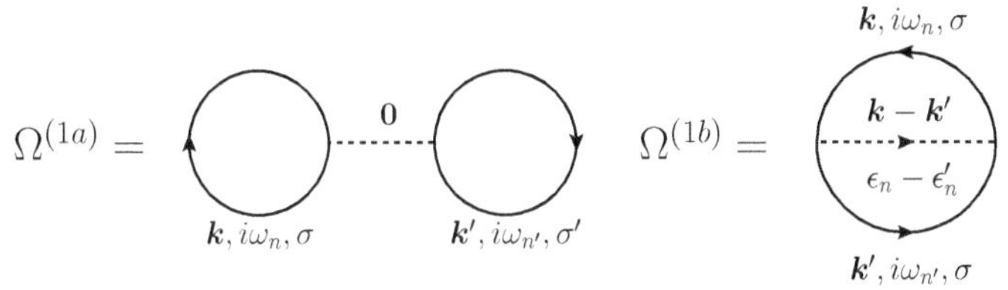

3.2 Green's Function Method for Many-Body Problem

Fig. 3.1 Diagrammatic representation of the first order contribution $\Omega^{(1)}$

$$\Omega^{(1b)} = -\frac{(k_B T)^2}{2} \sum_{kk'} \sum_{\epsilon_n \epsilon_{n'}} \sum_{\sigma} V(k-k') G_{\sigma}^{(0)}(k) \exp(i\epsilon_n \delta) G_{\sigma}^{(0)}(k') \exp(i\epsilon_{n'}\delta). \tag{3.41}$$

Here δ is an infinitesimal positive number. The corresponding diagrams are shown in Fig. 3.1. The rule to draw diagrams for the nth order contribution is as follows:

1. Assign $-k_B T G^{(0)}$ a solid line and the interaction $-V$ a broken line.
2. Assign variables so as to conserve four-momentum at each vertex.
3. Multiply a factor of $(-1)^{n_l}(-1)^n \frac{1}{2^n} \frac{1}{n!}(k_B T)^{n+1}$ and the number of equivalent diagrams. Here n_l is the number of loops made from solid lines.

As can be seen from the fact that a solid line corresponds to $G^{(0)}$, namely the contraction in the Bloch–de Dominicis decomposition, the connectivity of a perturbation term is identical to that of lines of a corresponding diagram. Therefore the above-mentioned linked-cluster theorem for the thermodynamic potential implies it is sufficient to collect the connected diagrams in the perturbation expansion. This diagrammatic representation is quite useful not only in the perturbation theory for the thermodynamic potential, but also that for the Green's function, which we shall describe in the next subsection.

3.2.5 Perturbation Theory for Green's Function

Similarly to the thermodynamic potential, here we describe how the Green's function is expressed in terms of the Feynman diagrams.

Using S defined above, Green's function can be expressed as

$$\left\langle T_u \left[\bar{c}_{k\sigma}(u) \bar{c}_{k\sigma}^{\dagger}(u') \right] \right\rangle = \frac{\text{Tr}\left\{\exp(-\beta H_0) T_u \left[S(\beta, u) c_{k\sigma}(u) S(u, u') c_{k\sigma}^{\dagger}(u') S(u', 0) \right]\right\}}{\text{Tr}\{\exp(-\beta H_0) S(\beta, 0)\}}$$

$$= \frac{\left\langle T_u \left[S(\beta, 0) c_{k\sigma}(u) c_{k\sigma}^{\dagger}(u') \right] \right\rangle_0}{\langle S(\beta, 0) \rangle_0}, \tag{3.42}$$

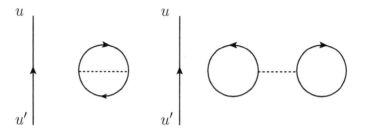

Fig. 3.2 First-order disconnected diagrams

Since the denominator has already been treated above, here we consider the numerator. For instance, the first-order contribution can be classified into a connected diagram (Fig. 3.3) and a disconnected one (Fig. 3.2). In these figures, $G^{(0)}$ lines going into (coming from) $u^{(\prime)}$ are called the external lines.

Similarly for the higher-order contribution can be expressed as the product of diagrams connected to the external lines and those disconnected. Since the sum of the contribution from the disconnected diagrams clearly coincides with $\langle S(\beta, 0)\rangle_0$, we have

$$G_\sigma(\boldsymbol{k}, u - u') = -\left\langle T_u \left[S(\beta, 0) c_{\boldsymbol{k}\sigma}(u) c_{\boldsymbol{k}\sigma}^\dagger(u') \right] \right\rangle_{0,c}. \tag{3.43}$$

Therefore, also for the Green's function, it is sufficient to collect the connected diagrams. In the Matsubara frequency domain, this becomes

$$G_\sigma(k) = -\int_0^\beta du \exp{(i\epsilon_n u)} \left\langle T_u \left[S(\beta, 0) c_{\boldsymbol{k}\sigma}(u) c_{\boldsymbol{k}\sigma}^\dagger(0) \right] \right\rangle_{0,c}. \tag{3.44}$$

Similarly to the thermodynamic potential, the Feynman rules for the perturbation expansion of $G_\sigma(k)$ are as follows:

1. Assign $-k_B T G^{(0)}$ a solid line and the interaction $-V$ a broken line.
2. Assign variables so as to conserve four-momentum at each vertex.
3. Multiply a factor $(-1)^{n_l}$.
4. Sum over with respect to all the variables except for the external line.

As discussed so far, the perturbation theory for both the thermodynamic potential and the Green's function can be done in a similar manner using the Feynman diagrams. Then we will see how these are related to each other (Figs. 3.1 and 3.3).

It can be seen from Fig. 3.3 that the first-order contribution of the Green's function can be interpreted as a diagram obtained by cutting open one $G^{(0)}$ line in the first-order diagram for the thermodynamic potential. Hence let us define $\Sigma_{\boldsymbol{k}\sigma}^{(1)}$ as a diagram obtained by removing one $G^{(0)}$ line in the first-order diagram for the thermodynamic potential. Using this quantity, we have

3.2 Green's Function Method for Many-Body Problem

Fig. 3.3 First-order connected diagrams

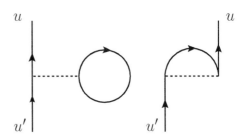

$$G_\sigma^{(1)}(k) = G_\sigma^{(0)}(k)\Sigma_\sigma^{(1)}(k)G_\sigma^{(0)}(k), \qquad (3.45)$$

$\Sigma_\sigma^{(1)}$ is the functional derivative of $\Omega^{(1)}$ with respect to $G^{(0)}$, which is given by

$$\Sigma_\sigma^{(1)}(k) = k_B T \sum_{k',n',\sigma'} V(0)G_{\sigma'}^{(0)}(k')\exp(i\epsilon_{n'}\delta) - k_B T \sum_{k',n'} V(k-k')G_\sigma^{(0)}(k')\exp(i\epsilon_{n'}\delta). \qquad (3.46)$$

We can also define $\Sigma^{(n)}$ for the general nth order terms via the functional derivative of $\Omega^{(n)}$.

$$G_\sigma^{(n)}(k) = G_\sigma^{(0)}(k)\Sigma_\sigma^{(n)}(k)G_\sigma^{(0)}(k), \qquad (3.47)$$

$$\Sigma_\sigma^{(n)}(k) = \beta\frac{\delta\Omega^{(n)}}{\delta G_\sigma^{(0)}(k)}. \qquad (3.48)$$

Let us look into more details of the diagrammatic structure of $\Sigma^{(n)}$. For a simple notation here we omit k and σ. Let us introduce the concept of a proper diagram, which can not be separated into two pieces by cutting a single $G^{(0)}$ line. By definition $\Sigma^{(n)}$ is a sum of the proper $\Sigma^{(n;p)}$ and improper part $\Sigma^{(n;ip)}$:

$$\Sigma^{(n)} = \Sigma^{(n;p)} + \Sigma^{(n;ip)}. \qquad (3.49)$$

Also by definition, an improper diagram is given by repetition of proper diagrams connected by $G^{(0)}$ lines. By introducing a sum of proper diagrams Σ,

$$\Sigma = \sum_{n=1}^{\infty} \Sigma^{(n;p)}, \qquad (3.50)$$

the Green's function is represented as

$$G = G^{(0)} + G^{(0)}\sum_{n=1}^{\infty}\left(\Sigma^{(n;p)} + \Sigma^{(n;ip)}\right)G^{(0)}$$
$$= G^{(0)} + G^{(0)}\Sigma G, \qquad (3.51)$$

Fig. 3.4 A diagrammatic representation of the Dyson equation

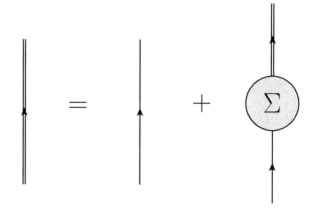

or

$$G_{k\sigma}(i\epsilon_n) = \left[\{G_\sigma^{(0)}(k)\}^{-1} - \Sigma_\sigma(k)\right]^{-1}$$
$$= [i\epsilon_n - \xi(k) - \Sigma_\sigma(k)]^{-1}, \quad (3.52)$$

This equation is called the Dyson equation, and Σ the self-energy. In Fig. 3.4, we show a diagrammatic representation of Eq. (3.51).

3.2.6 Luttinger–Ward Functional

As we shall see here, a concept of the self-energy is useful not only for the Green's function but also for the thermodynamic potential. Let us introduce a concept of an irreducible diagram, which can not be generated by replacing $G^{(0)}$ by G (full Green's function) in any lower-order diagram. By definition, the self-energy is obtained by summing up all the irreducible diagrams in which $G^{(0)}$ lines are replaced by G:

$$\Sigma_\sigma(k) = \sum_{n=1}^{\infty} \Sigma_\sigma^{(n;\text{irr})}(k; [G]). \quad (3.53)$$

Here the superscript $(n; \text{irr})$ means the contribution of the nth order irreducible diagrams and the argument $[G]$ is to emphasize the replacement of $G^{(0)}$ by G.

To obtain further systematics for collecting diagrams, let us define the "generating functional" $\Phi[G]$ of the self-energy by the functional derivative:

$$\frac{\delta \Phi[G]}{\delta G} = \Sigma_\sigma(k), \quad (3.54)$$

3.2 Green's Function Method for Many-Body Problem

or more explicitly:

$$\Phi[G] = k_B T \sum_{\sigma,k,n} \frac{1}{2n} G_\sigma(k) \Sigma_\sigma^{(n;irr)}(k;[G]). \quad (3.55)$$

Considering an adiabatic connection from a non-interacting system to a fully interacting one, one can obtain the following expression for the thermodynamic potential:

$$\Omega = -k_B T \sum_{\epsilon_n, k\sigma} \exp(i\epsilon_n \delta) \left[\ln\left(-\{G_\sigma(k)\}^{-1}\right) + G_\sigma(k)\Sigma_\sigma(k) \right] + \Phi[G], \quad (3.56)$$

This $\Phi[G]$ is called the Luttinger–Ward functional for the thermodynamic potential [5]. Note that $\Omega[\Sigma]$ as a functional of the true self-energy has a variational feature since $\delta\Omega/\delta\Sigma = 0$.

Baym-Kadanoff Scheme for Conserving Approximation

Here we comment on a prominent advantage of the Luttinger–Ward formalism in constructing an approximated self-energy. According to the Baym–Kadanoff scheme [6], it is automatically guaranteed that the correlation functions obey conservation laws if an approximated form is written down in the level of the Luttinger–Ward functional $\Phi[G]$. This enables us to systematically construct a conserving approximation with the above-mentioned features of the Luttinger–Ward formalism.

3.2.7 BCS Theory

Conventional superconductivity is well understood based on the microscopic theory proposed by Bardeen, Cooper, and Schrieffer [7], which is called the BCS theory. In the BCS theory, superconductivity arise from an attractive electron–electron interaction mediated by the electron-phonon coupling.

Firstly, let us see how the attractive interaction is obtained. Here we consider the perturbation process of the Hamiltonian of the electron-phonon coupling:

$$\mathcal{H}_{\text{el-ph}} = \frac{1}{\sqrt{N}} \sum_{k,\sigma} \alpha(\boldsymbol{q})(a_{\boldsymbol{q}} + a^\dagger_{-\boldsymbol{q}}) c^\dagger_{k\sigma} c_{k\sigma}. \quad (3.57)$$

To note that N denotes the number of the atomic sites and hence the number of k-points since we consider the tight-binding model here. Here $a^\dagger$ and a are the creation and the annihilation operator of a phonon, respectively. The first-order process corresponds to the absorption (emission) of a phonon with momentum $\boldsymbol{q}(-\boldsymbol{q})$. Here we consider a second order process which has the same initial and final state, using the excited state due to this process as an intermediate state. This second order process yields the energy gain of $|\alpha(q)|/2\omega(q)$ with the transition amplitude $\alpha(q)$ and the energy deference $\omega(q)$. Considering this, as an effective electron–electron interaction, we obtain

$$\mathcal{H}_{\text{eff,el}} = -\frac{2}{N} \sum_{\substack{k,k',q \\ \sigma\sigma'}} \frac{|\alpha(q)|^2}{\omega(q)} c^{\dagger}_{k'+q\sigma} c^{\dagger}_{k-q\sigma'} c_{k'\sigma'} c_{k\sigma}, \qquad (3.58)$$

which is attractive as can be seen from the negative sign.

Given this, we take the effective Hamiltonian for electrons as

$$\mathcal{H} = \sum_{k,\sigma} \xi(k) c^{\dagger}_{k\sigma} c_{k\sigma} + \sum_{k,k'} V(k-k') c^{\dagger}_{k'+q\sigma} c^{\dagger}_{k-q\sigma'} c_{k'\sigma'} c_{k\sigma}. \qquad (3.59)$$

Note that the interaction V, which can be attractive, includes both the attractive interaction mediated by phonons and the repulsive interaction. Here we restrict the interaction term so as to works only for pairs of electrons with opposite spins (spin singlet) and with total momentum of $q = 0$:

$$\mathcal{H} = \sum_{k,\sigma} \xi(k) c^{\dagger}_{k\sigma} c_{k\sigma} + \sum_{k,k'} V(k-k') c^{\dagger}_{-k'\downarrow} c^{\dagger}_{k'\uparrow} c_{k\uparrow} c_{-k\downarrow}. \qquad (3.60)$$

Since this Hamiltonian contains the two-body interaction, it is extremely difficult to solve analytically. Here we take a mean-field approximation by introducing $\langle c_{k\uparrow} c_{-k\downarrow} \rangle$ as the order parameter of superconductivity:

$$\mathcal{H}_{\text{BCS}} = \sum_{k,\sigma} \xi(k) c^{\dagger}_{k\sigma} c_{k\sigma} - \sum_k \Delta(k) c^{\dagger}_{-k\downarrow} c^{\dagger}_{k\uparrow} - \sum_k \Delta(k)^* c_{k\uparrow} c_{-k\downarrow}, \qquad (3.61)$$

where

$$\Delta(k) = -\sum_k V(k-k') \langle c'_{k\uparrow} c'_{-k\downarrow} \rangle \qquad (3.62)$$

is called the gap function.

Since Eq. (3.61) is written in a bilinear form, by introducing the unitary transformation which mixes an electron and a hole

$$\begin{pmatrix} \alpha_{k\uparrow} \\ \alpha^{\dagger}_{-k\downarrow} \end{pmatrix} = \begin{pmatrix} u_k & -v_k \\ -v_k^* & u_k \end{pmatrix} \begin{pmatrix} c_{k\uparrow} \\ c^{\dagger}_{-k\downarrow} \end{pmatrix}, \qquad (3.63)$$

we can choose u_k and v_k ($|u_k|^2 + |v_k|^2 = 1$) so that this transformation diagonalizes Eq. (3.61):

3.2 Green's Function Method for Many-Body Problem

$$\mathcal{H}_{BCS} = E_{GS} + \sum_k E(k) \left(\alpha_{k\uparrow}^\dagger \alpha_{k\uparrow} + \alpha_{-k\downarrow}^\dagger \alpha_{-k\downarrow} \right), \quad (3.64)$$

$$E(k) = \pm\sqrt{\xi(k)^2 + |\Delta(k)|^2} \quad (3.65)$$

Here E_{GS} is the energy of the state without quasiparticles described by α. Note that one can easily show $\alpha^{(\dagger)}$ is the annihilation (creation) operator of fermions. $E(k)$ corresponds to the quasiparticle excitation, and the energy of $|\Delta(k)|$ is necessary to excite a quasiparticle.

The gap function $\Delta(k)$ is determined by the self-consistent equation called the gap equation:

$$\Delta(k) = -\sum_{k'} V(k - k') \frac{\Delta(k')}{2E(k')} \tanh\left(\frac{1}{2}\beta E(k')\right) \quad (3.66)$$

Assuming a constant $\Delta(k) = \Delta$, $\Delta(k) = 0$ for repulsive interactions and $\Delta(k)$ can be a nontrivial solution for attractive interactions. Even for repulsive interactions, $\Delta(k)$ can be finite if $\Delta(k)$ reverses its sign in the Brillouin zone.

3.2.8 Eliashberg Equation

In the BCS theory, the attractive interaction is derived by assuming the electron-phonon coupling to be weak. Here we introduce the Eliashberg theory for systems with a strong electron-phonon coupling and derive the Eliashberg equation corresponding to the gap equation in the BCS theory.

Let us consider the Hamiltonian explicitly involving the electron-phonon coupling:

$$\mathcal{H} = \sum_{k,\sigma} \xi(k) c_{k\sigma}^\dagger c_{k\sigma} + \sum_q \xi(q) a_q^\dagger a_q + \frac{1}{\sqrt{N}} \sum_{k,\sigma} \alpha(q)(a_q + a_{-q}^\dagger) c_{k\sigma}^\dagger c_{k\sigma}, \quad (3.67)$$

Here $\omega(q)$ is the energy dispersion of a phonon. The equation of motion for the Green's function can be written as

$$-\left[\frac{\partial}{\partial u} + \xi(k)\right] G(k, u - u') = \delta(u - u') + \frac{1}{\sqrt{N}} \sum_{k'} \alpha(k') \Gamma(k', k, u, u', u'').$$

$$(3.68)$$

Here Γ is defined as

$$\Gamma(k', k, u, u', u'') = \left\langle T_u \phi_{k'}(u) c_{k+k'\uparrow}(u'') c_{k\uparrow}^\dagger(u') \right\rangle, \quad (3.69)$$

$$\phi_k(u) \equiv a_k^\dagger + a_{-k}. \quad (3.70)$$

Using the equation of motion for a, one can obtain

$$\left[\frac{\partial^2}{\partial u^2} - \omega(k')\right]\Gamma(k', k, u, u', u'') = -\frac{2\omega(k')\alpha(k')}{\sqrt{N}} \sum_{q,\sigma} \left\langle T_u c_{q\sigma}^\dagger(u) c_{q-k'\sigma}(u) c_{k+k'\uparrow}(u'') c_{k\uparrow}^\dagger(u') \right\rangle, \quad (3.71)$$

Since the Green's function of a phonon is defined by

$$D(k, u - u') = -\left\langle T\phi_k(u)\phi_{-k}(u')\right\rangle, \quad (3.72)$$

one can obtain the equation of motion for the non-interacting phonon:

$$\left[\frac{\partial^2}{\partial u^2} - \omega(k)\right] D(k, u - u') = 2\omega(k)\delta(u - u'). \quad (3.73)$$

Equations (3.71) and (3.73) result in

$$\Gamma(k', k, u, u', u'') = -\int_0^\beta du_1 \frac{\alpha(k')}{\sqrt{N}} D(k', u - u_1)$$
$$\times \sum_{q,\sigma} \left\langle T_u c_{q\sigma}^\dagger(u_1) c_{q-k'\sigma}(u_1) c_{k+k'\uparrow}(u) c_{k\uparrow}^\dagger(u') \right\rangle. \quad (3.74)$$

Substituting this into Eq. (3.68), we obtain

$$-\left[\frac{\partial}{\partial u} + \xi(k)\right] G(k, u - u') = \delta(u - u') - \frac{1}{N}\int_0^\beta du_1 \alpha^2(k') D(k', u - u_1)$$
$$\times \sum_{q,\sigma} \left\langle T_u c_{q\sigma}^\dagger(u_1) c_{q-k'\sigma}(u_1) c_{k+k'\uparrow}(u) c_{k\uparrow}^\dagger(u') \right\rangle. \quad (3.75)$$

Let us define the anomalous Green's function as

$$F^*(k, u - u') = -\left\langle T c_{-k\downarrow}^\dagger(u) c_{k\uparrow}^\dagger(u') \right\rangle, \quad (3.76)$$

and we employ the mean-field approximation for Eq. 3.75 as

$$\sum_{q,\sigma} \left\langle T_u c_{q\sigma}^\dagger(u) c_{q-k'\sigma}(u) c_{k+k'\uparrow}(u'') c_{k\uparrow}^\dagger(u') \right\rangle$$
$$\simeq \left\langle T_u c_{k+k'\uparrow}(u) c_{k+k'\uparrow}^\dagger(u_1) \right\rangle \left\langle T_u c_{k\uparrow}(u_1) c_{k\uparrow}^\dagger(u') \right\rangle$$
$$- \left\langle T_u c_{k+k'\uparrow}(u) c_{-k-k'\downarrow}(u_1) \right\rangle \left\langle T_u c_{-k\downarrow}^\dagger(u_1) c_{k\uparrow}^\dagger(u') \right\rangle. \quad (3.77)$$

3.2 Green's Function Method for Many-Body Problem

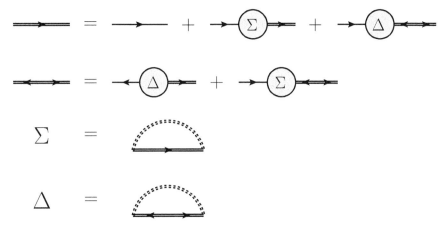

Fig. 3.5 Diagrammatic expression of the Eliashberg equation. Double-headed arrow means the anomalous Green's function

Similarly to this, one can derive the equation of motion for the anomalous Green's function.

Fourier transforming the obtained equations with respect to $u - u'$,

$$[i\epsilon_n - \xi(\mathbf{k}) - \Sigma(\mathbf{k}, i\epsilon_n)] G(\mathbf{k}, i\epsilon_n) - \Delta(\mathbf{k}, i\epsilon_n) F^*(\mathbf{k}, i\epsilon_n) = 1, \quad (3.78)$$

$$[i\epsilon_n - \xi(\mathbf{k}) - \Sigma(\mathbf{k}, -i\epsilon_n)] F^*(\mathbf{k}, i\epsilon_n) - \Delta^*(\mathbf{k}, i\epsilon_n) G(\mathbf{k}, i\epsilon_n) = 0. \quad (3.79)$$

Here Σ and Δ are the (normal) and the anomalous self-energies, respectively:

$$\Sigma(\mathbf{k}, i\epsilon_n) = -\frac{k_B T}{N} \sum_{\mathbf{q}, m} \alpha^2(\mathbf{k} - \mathbf{q}) D(\mathbf{k} - \mathbf{q}, i\epsilon_n - i\epsilon_m) G(\mathbf{q}, i\epsilon_m), \quad (3.80)$$

$$\Delta(\mathbf{k}, i\epsilon_n) = \frac{k_B T}{N} \sum_{\mathbf{q}, m} \alpha^2(\mathbf{k} - \mathbf{q}) D(\mathbf{k} - \mathbf{q}, i\epsilon_n - i\epsilon_m) F(\mathbf{q}, i\epsilon_m), \quad (3.81)$$

These simultaneous equations are called the Eliashberg equation, which is a generalization of the Dyson equation for superconductivity. More precisely, we should replace the non-interacting phonon Green's function by the full phonon Green's function. The diagrammatic expression of the Eliashberg equation is shown in Fig. 3.5.

Near the transition temperature T_c, one can linearize the Eliashberg equation by assuming F and Δ are small.

$$\lambda \Delta(k) = -\frac{k_B T}{N} \sum_{k'} \alpha^2(\mathbf{k} - \mathbf{k'}) D(k - k') G(k') G(-k') \Delta(k') \quad (3.82)$$

Here we introduce λ in order to solve this as an eigenvalue problem numerically and hence $\lambda = 1$ at $T = T_c$. In many cases, it is computationally demanding to perform the actual calculations at low temperature where λ reaches unity, so in the present study, we adopt λ, obtained at a fixed temperature, to measure how close the system is to superconductivity. In a strict sense, Δ in the *linearized* Eliashberg equation can not be called as a self-energy due to the linearization, but still we will simply call it the anomalous self-energy if there would be no confusion.

Here we have treated the electron-phonon system, however it is possible to extend this formalism to superconductivity mediated by the electron correlation. Defining the anomalous self-energy as in the electron-phonon system, and introducing the matrix (Nambu) representation of the Green's function as

$$\hat{G} = \begin{pmatrix} G(k) & F(k) \\ F^*(k) & G(-k) \end{pmatrix} \quad (3.83)$$

and the corresponding self-energy, we can see that the Dyson equation in this representation is essentially the same as the Eliashberg equation.

3.3 Perturbative Approximations for Green's Function

In this section, we describe approximations relevant to the present study based on the perturbation theory described above. First, we introduce the spin and charge susceptibilities which are building blocks for the perturbative approximations in systems with spin and charge fluctuations. Second, we summarize the random phase approximation (RPA) which is one of the most simple approximations. Finally we describe a widely used approximation, the fluctuation exchange (FLEX) approximation [8, 9], which is a generalization of RPA and conserving in the Baym–Kadanoff sense.

3.3.1 Spin/Charge Susceptibility

The spin density operator S_q and the charge density operator ρ_q are defined as follows in the second quantization representation:

$$S_q = \frac{1}{2} \sum_k \sum_{\alpha,\beta} c_{k\alpha}^\dagger \sigma_{\alpha\beta} c_{k+q\beta}, \quad (3.84)$$

$$\rho_q = \sum_k \left(c_{k\uparrow}^\dagger c_{k+q\uparrow} + c_{k\downarrow}^\dagger c_{k+q\downarrow} \right). \quad (3.85)$$

Here σ is the Pauli matrices. Transforming x and y components into $\pm$, we have

$$S_q^+ = \sum_k c_{k\uparrow}^\dagger c_{k+q\downarrow}, \quad (3.86)$$

$$S_q^- = \sum_k c_{k-q\downarrow}^\dagger c_{k\uparrow} = S_{-q}^{+\dagger}. \quad (3.87)$$

3.3 Perturbative Approximations for Green's Function

With the aid of the linear response theory, we can write down the transverse $\chi_s^{\pm}$ and the longitudinal spin susceptibility χ_s^{zz} as well as the charge susceptibility χ_c as

$$\chi_s^{\pm}(k) = \int_0^{\beta} du \exp(i\omega_m u) \frac{1}{N} \langle \bar{S}_k^+(u) \bar{S}_{-k}^-(0) \rangle, \tag{3.88}$$

$$\chi_s^{zz}(k) = \int_0^{\beta} du \exp(i\omega_m u) \frac{1}{N} \langle \bar{S}_k^z(u) \bar{S}_{-k}^z(0) \rangle$$

$$= \frac{1}{4} \{ \chi^{\uparrow\uparrow}(k) + \chi^{\downarrow\downarrow}(k) - \chi^{\uparrow\downarrow}(k) - \chi^{\downarrow\uparrow}(k) \}, \tag{3.89}$$

$$\chi_c(k) = \int_0^{\beta} du \exp(i\omega_m u) \frac{1}{2N} \langle \bar{\rho}_k(u) \bar{\rho}_{-k}(0) \rangle$$

$$= \frac{1}{2} \{ \chi^{\uparrow\uparrow}(k) + \chi^{\downarrow\downarrow}(k) + \chi^{\uparrow\downarrow}(k) + \chi^{\downarrow\uparrow}(k) \}. \tag{3.90}$$

Here ω_m is the bosonic Matsubara frequency and $\chi^{\sigma\sigma'}(q)$ is defined as

$$\chi^{\sigma\sigma'}(q) = \int_0^{\beta} du \exp(i\omega_m u) \frac{1}{N} \sum_{k,l} \langle \bar{c}_{k\sigma}^{\dagger}(u) \bar{c}_{k+q\sigma}(u) \bar{c}_{l+q\sigma'}^{\dagger}(0) \bar{c}_{l\sigma'}(0) \rangle, \tag{3.91}$$

Assuming a paramagnetic state, we have the relationships $\chi^{\uparrow\downarrow} = \chi^{\downarrow\uparrow}$ and $\chi^{\uparrow\uparrow} = \chi^{\downarrow\downarrow}$ by the spin SU(2) symmetry. Then we can reduce the spin susceptibilities to a single quantity:

$$\chi_s^{\pm}(k) = 2\chi_s^{zz}(k) \equiv \chi_s(k). \tag{3.92}$$

3.3.2 Random Phase Approximation (RPA)

In order to obtain the spin/charge susceptibility, it is necessary to calculate $\chi^{\sigma\sigma'}$. Although the perturbation theory can be formulated for two-particle quantities similarly to the case for the single-particle Green's function, an approximation is necessary since it is virtually impossible to collect all the diagrams just as in the single-particle case. Here we will see how the susceptibilities are expressed within the RPA. Hereafter, we consider the multi-orbital Hubbard model defined by Eq. (3.12).

In the multi-orbital systems, the Green's function is defined just as in the single-orbital system by considering the orbital degrees of freedom:

$$G_{\alpha\beta}(k,u) = \langle T_u [\bar{c}_{k\sigma}^{\alpha}(u) c_{k\sigma}^{\beta\dagger}] \rangle. \tag{3.93}$$

The perturbation theory is also formulated in the same manner using the matrix representation. The non-interacting Green's function is given by

$$\hat{G}^{(0)}(k)^{-1} = \left[i\omega_n \hat{I} - (\hat{\mathcal{H}}(k) - \mu \hat{I})\right]. \tag{3.94}$$

Here I is an identity matrix. The Dyson equation also becomes

$$\hat{G}(k)^{-1} = \hat{G}^{(0)}(k)^{-1} - \hat{\Sigma}(k). \tag{3.95}$$

In RPA, the so-called ladder and bubble type diagrams, whose diagrams for the single-band model are shown for instance in Fig. 3.6, are taken into account. Let us introduce the interaction vertices in the spin $\hat{S}$ and charge channel $\hat{C}$ for the multi-orbital model as

$$S_{\alpha\beta\gamma\delta} = \begin{cases} U & (\alpha = \beta = \gamma = \delta) \\ U' & (\alpha = \gamma \neq \beta = \delta) \\ J & (\alpha = \beta \neq \gamma = \delta) \\ J' & (\alpha = \delta \neq \beta = \gamma) \end{cases}, \tag{3.96}$$

$$C_{\alpha\beta\gamma\delta} = \begin{cases} U & (\alpha = \beta = \gamma = \delta) \\ 2J - U' & (\alpha = \gamma \neq \beta = \delta) \\ 2U' - J & (\alpha = \beta \neq \gamma = \delta) \\ J' & (\alpha = \delta \neq \beta = \gamma) \end{cases}. \tag{3.97}$$

Defining the interaction between electrons with the same (different) spin $V_{\sigma\sigma(\sigma\bar{\sigma})}$ as

$$V_{\sigma\sigma} = \frac{1}{2}\left(\hat{C} - \hat{S}\right), \tag{3.98}$$

$$V_{\sigma\bar{\sigma}} = \frac{1}{2}\left(\hat{C} + \hat{S}\right), \tag{3.99}$$

we have

$$\hat{\chi}^{\sigma\sigma}(q) = \hat{\chi}^{\text{irr}}(q) - \hat{\chi}^{\text{irr}}(q)V_{\sigma\sigma}\hat{\chi}^{\sigma\sigma}(q) - \hat{\chi}^{\text{irr}}(q)V_{\sigma\bar{\sigma}}\hat{\chi}^{\bar{\sigma}\sigma}(q), \tag{3.100}$$

$$\hat{\chi}^{\sigma\bar{\sigma}}(q) = -\hat{\chi}^{\text{irr}}(q)V_{\sigma\bar{\sigma}}\hat{\chi}^{\bar{\sigma}\bar{\sigma}}(q) - \hat{\chi}^{\text{irr}}(q)V_{\sigma\sigma}\hat{\chi}^{\sigma\bar{\sigma}}(q), \tag{3.101}$$

where the irreducible susceptibility $\hat{\chi}^{\text{irr}}(q)$ is defined as the non-interacting susceptibility:

$$\chi^{\text{irr}}_{l_1 l_2, l_3 l_4}(q) \equiv -\frac{T}{N} \sum_k G^{(0)}_{l_3 l_1}(k) G^{(0)}_{l_2 l_4}(k+q). \tag{3.102}$$

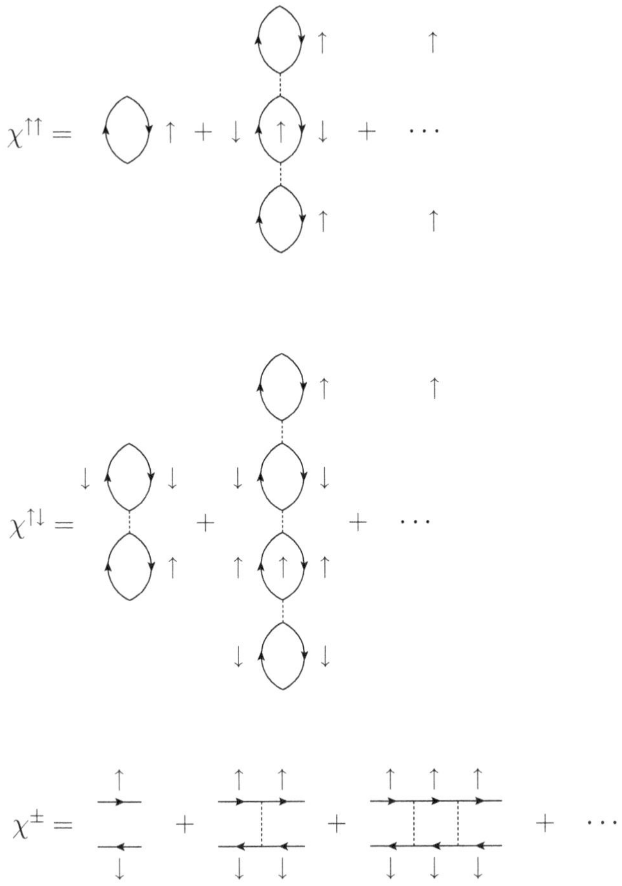

3.3 Perturbative Approximations for Green's Function

Fig. 3.6 Feynman diagrams of the susceptibilities within RPA for the single-orbital Hubbard model

Therefore the spin and charge susceptibilities can be expressed as

$$\hat{\chi}_s(q) = \left(\hat{I} - \hat{\chi}^{\text{irr}}(q)\hat{S}\right)^{-1} \hat{\chi}^{\text{irr}}(q), \quad (3.103)$$

$$\hat{\chi}_c(q) = \left(\hat{I} + \hat{\chi}^{\text{irr}}(q)\hat{C}\right)^{-1} \hat{\chi}^{\text{irr}}(q). \quad (3.104)$$

3.3.3 Fluctuation Exchange (FLEX) Approximation

Since RPA neglects the couplings between fluctuations with different q's (mode-mode couplings), it is insufficient for systems with strong fluctuations. The FLEX approximation [8, 9] can be regarded as a diagrammatic generalization of RPA, which self-consistently treats strong fluctuations. In the FLEX approximation, the irreducible susceptibility consists of the *renormalized* Green's function $G(k)$:

$$\chi^{\text{irr}}_{l_1 l_2, l_3 l_4}(q) \equiv -\frac{T}{N} \sum_k G_{l_3 l_1}(k) G_{l_2 l_4}(k+q). \quad (3.105)$$

The effective electron–electron interaction Γ for obtaining the self-energies calculated by collecting bubble- and ladder-type diagrams:

$$\Gamma(q) = \frac{3}{2}\hat{S}\hat{\chi}_s(q)\hat{S} + \frac{1}{2}\hat{C}\hat{\chi}_c(q)\hat{C} - \frac{1}{4}\left(\hat{C}+\hat{S}\right)\chi^{\text{irr}}(q)\left(\hat{C}+\hat{S}\right) + \frac{3}{2}\hat{S} - \frac{1}{2}\hat{C}. \quad (3.106)$$

The Green's function is obtained by solving the Dyson equation

$$\hat{G}(k)^{-1} = \hat{G}^{(0)}(k)^{-1} - \hat{\Sigma}(k) \quad (3.107)$$

$$\Sigma_{ll'}(k) = \frac{1}{2}\frac{k_B T}{N}\sum_q \Gamma_{lml'm'}(q) G_{mm'}(k-q) \quad (3.108)$$

using the obtained effective interaction. Solving these equations self-consistently, one can obtain the renormalized Green's function within the FLEX approximation.

In this approximation, we consider bubble- and ladder-type diagrams consisting of the renormalized Green's function. These diagrams, in fact, can be generated by the Luttinger–Ward functional considering bubble- and ladder-type diagrams (Fig. 3.7) and hence this method guarantees the FLEX approximation to be conserving by the Baym–Kadanoff scheme.

Since the Luttinger–Ward formalism can be extended to the functional containing anomalous quantities, the anomalous self-energy Δ can be derived by the functional derivative with respect to F. Therefore we obtain the singlet pairing interaction $\Gamma^s(q)$ mediated mainly by spin fluctuations, which is plugged into the linearized Eliashberg equation for superconductivity,

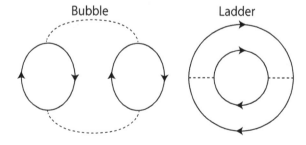

Fig. 3.7 Example for bubble- and ladder-type diagrams of the Luttinger–Ward functional considered in the FLEX approximation

3.3 Perturbative Approximations for Green's Function

$$\lambda \Delta_{ll'}(k) = -\frac{T}{N} \sum_{k'm_i} \Gamma^s_{lm_1m_4l'}(k-k') G_{m_1m_2}(k') \Delta_{m_2m_3}(k') G_{m_4m_3}(-k'), \quad (3.109)$$

$$\Gamma^s(q) = \frac{3}{2}\hat{S}\hat{\chi}_s(q)\hat{S} - \frac{1}{2}\hat{C}\hat{\chi}_c(q)\hat{C} + \frac{1}{2}\left(\hat{S}+\hat{C}\right). \quad (3.110)$$

In the present study, the largest positive eigenvalue of the equation is obtained with the power method. First, we calculate the eigenvalue λ_1, which may be negative. Then, we obtain the eigenvalue λ_2 by solving the same problem with the shifted power method with the shift of $|\lambda_1|$. This ensures $\lambda_2 - |\lambda_1|$ is the largest positive eigenvalue of the equation.

3.4 Dynamical Mean-Field Theory and Its Combination with FLEX

In this section, firstly, we describe the dynamical mean-field theory (DMFT) [10–12]. It is capable of treating strong local correlation effects such as the Mott transition and the physics in the vicinity of it, which can not be captured by the above-mentioned RPA and FLEX approximation.

However, since DMFT neglects the spatial correlation exemplified by magnetic/charge fluctuations, it has to be generalized in order to treat superconductivity mediated by spin fluctuation. To treat non-local correlation effects, various extensions of DMFT incorporating spatial fluctuations have been proposed. To include the momentum dependence of the self-energy, there are roughly two families of generalizations: cluster extensions and diagrammatic ones. In the former family, the lattice system is mapped onto a small-size cluster impurity problem in momentum space (DCA; dynamical cluster approximation) [13] or real space (cDMFT ; cluster/cellular DMFT) [14, 15]. In the latter family, non-local correlation is diagrammatically treated starting from DMFT such as the dynamical vertex approximation [16], the dual fermion method [17–19], and the functional renormalization group combined with DMFT [20].

Along this line, we will introduce the combination of DMFT and the FLEX approximation which is a diagrammatic extension of DMFT [21–23]. This method is computationally cheap in comparison to other generalizations and hence enables us to analyze multi-band systems in a wide range of parameters [13–20].

3.4.1 Path-Integral Representation

For later convenience, we briefly introduce the path integral representation for the partition function and the Green's function. For more details we refer to Ref. [24]. Let us define the Grassmann numbers $c^{(\dagger)}$ as the eigenvalue of the creation(annihilation) operator satisfying the anti-commutation relation:

$$\{c_1, c_2\} = 0, \tag{3.111}$$

$$c^2 = 0. \tag{3.112}$$

Note that c and $c^\dagger$ are independent.

Then, one can write down the grand partition function Ξ and the Green's function for the fermionic Hamiltonian $\mathcal{H}(\{c\}, \{c^\dagger\})$ in terms of functional integrals over Grassmann numbers:

$$\Xi = \text{Tr}\left\{e^{-\beta(\mathcal{H}-\mu N)}\right\} = \int \mathcal{D}c^\dagger \mathcal{D}c \exp(-\mathcal{S}) \tag{3.113}$$

$$G_{\alpha\beta}(u) = \frac{1}{Z}\int \mathcal{D}c^\dagger \mathcal{D}c\, c_\alpha(u) c_\beta^\dagger(0) \exp(-\mathcal{S}) \tag{3.114}$$

with the action

$$\mathcal{S} = \int_0^\beta du \left[\sum_\alpha c_\alpha^\dagger (\partial_u - \mu) c_\alpha + \mathcal{H}(\{c\}, \{c^\dagger\})\right], \tag{3.115}$$

where antiperiodic boundary conditions are imposed on $c_\alpha(u)$. Note that the above path-integral expressions are just shorthand notations for the discretized imaginary time u in the numerical calculations.

3.4.2 Dynamical Mean-Field Theory (DMFT)

Let us consider the single-band Hubbard model, for simplicity. The generalization to a multi-band system can be done straightforwardly. In DMFT, a lattice problem is solved by mapping it onto a single impurity problem. First, we derive an effective action for the Anderson impurity model. And then, we will show that an effective action for the Hubbard model is equivalent to that for the impurity model within a certain approximation so that the self-energy does not depend on the momentum.

The Hamiltonian of the Anderson model is given by

$$\mathcal{H} = \sum_{k\sigma} \epsilon(k) c_{k\sigma}^\dagger c_{k\sigma} + \sum_{k\sigma}\left(V_{kf} c_{k\sigma}^\dagger f_\sigma + \text{h.c.}\right) + \epsilon_f \sum_\sigma f_\sigma^\dagger f_\sigma + U n_{f\uparrow} n_{f\downarrow}, \tag{3.116}$$

where, $c_{k\sigma}^{(\dagger)}$ and $f_\sigma^{(\dagger)}$ are the annihilation (creation) operators for a conduction electron and an impurity, respectively. ϵ_k is the energy of a conduction electron, ϵ_f that of the f-level, U the interaction, and V_{kf} the hybridization between a conduction electron and an f electron. The action of this model can be written as

3.4 Dynamical Mean-Field Theory and Its Combination with FLEX

$$S = \sum_{nk\sigma} \left[(-i\omega_n + \epsilon_k) c_{k\sigma}^\dagger(i\omega_n) c_{k\sigma}(i\omega_n) + V_{kf} c_{k\sigma}^\dagger(i\omega_n) f_\sigma(i\omega_n) + V_{kf}^* f_\sigma^\dagger(i\omega_n) c_{k\sigma}(i\omega_n) \right]$$

$$+ \sum_{n\sigma} (i\omega_n + \epsilon_k) f_\sigma^\dagger(i\omega_n) f_\sigma(i\omega_n) + U \int_0^\beta n_{f\uparrow} n_{f\downarrow}, \quad (3.117)$$

where $c_{k\sigma}$, $c_{k\sigma}^\dagger$, f_σ, and $f_\sigma^\dagger$ are Grassmann numbers corresponding to the above operators. Since the action is bilinear in the conduction electrons, one can integrate them out by a Gaussian integral from the partition function:

$$Z = \int \mathcal{D}f \mathcal{D}f^\dagger \exp\left[-\sum_{n\sigma}(-i\omega_n + \epsilon_f + \Delta(i\omega_n)) f_\sigma^\dagger(i\omega_n) f_\sigma(i\omega_n) \right.$$

$$\left. - U \int_0^\beta n_{f\uparrow} n_{f\downarrow} + 2\sum_{nk} \ln(i\omega_n + \epsilon_k) \right]. \quad (3.118)$$

Here $\Delta(i\omega)$ is called the hybridization function defined by

$$\Delta(i\omega) = \sum_k \frac{|V_{kf}|^2}{i\omega_n - \epsilon_k} \quad (3.119)$$

The last term in the exponential can be omitted in calculations of quantities not containing c and $c^\dagger$. Therefore the effective action $\mathcal{S}_{\text{eff}}$ reads

$$\mathcal{S}_{\text{eff}} = \sum_{n\sigma} \left[-i\omega_n + \epsilon_f + \Delta(i\omega_n) \right] f_\sigma^\dagger(i\omega_n) f_\sigma(i\omega_n) + U \int_0^\beta n_{f\uparrow} n_{f\downarrow}$$

$$= \sum_{n\sigma} f_\sigma^\dagger(i\omega_n) G_f^{(0)}(i\omega_n)^{-1} f_\sigma(i\omega_n) + U \int_0^\beta n_{f\uparrow} n_{f\downarrow}, \quad (3.120)$$

where $G_f^{(0)}(i\omega_n) = \left[-i\omega_n + \epsilon_f + \Delta(i\omega_n) \right]^{-1}$ is the impurity Green's function in the non-interacting case.

Let us now consider the single band Hubbard model with the Hamiltonian,

$$\mathcal{H} = \sum_{i,j,\sigma} t_{ij} c_{i\sigma}^\dagger c_{j\sigma} + U \sum_i n_{i\uparrow} n_{i\downarrow} - \mu \sum_{i\sigma} n_{i\sigma}. \quad (3.121)$$

The partition function and the action can be expressed as

$$Z = \int \prod_{i,\sigma} \mathcal{D} c_{i,\sigma}^\dagger \mathcal{D} c_{i,\sigma} \exp(-\mathcal{S}) \quad (3.122)$$

$$\mathcal{S} = \int_0^\beta du \left\{ \sum_{i,j,\sigma} c_{i\sigma}^\dagger(u) \left[(\partial_u - \mu)\delta_{i,j} + t_{ij} \right] c_{j\sigma}(u) + U \sum_i n_{i\uparrow}(u) n_{i\downarrow}(u) \right\},$$
$$(3.123)$$

where we introduce the Grassmann numbers $c^\dagger$ and c corresponding to the creation and annihilation operators.

In the DMFT framework, we focus on a specific site (and call it site 0), and then integrate the contribution from the other sites to the action. To this end, let us decompose the action into three parts as follows:

$$S = S_0 + S_{0H} + S_H \tag{3.124}$$

$$S_0 = \int_0^\beta du \left[\sum_\sigma c_{0\sigma}^\dagger(u)(\partial_u - \mu)c_{j\sigma}(u) + U n_{0\uparrow} n_{0\downarrow} \right] \tag{3.125}$$

$$S_{0H} = \int_0^\beta du \left[\sum_{i,\sigma} \left[t_{i0} c_{i\sigma}^\dagger(u) c_{0\sigma}(u) + t_{0i} c_{0\sigma}^\dagger(u) c_{i\sigma}(u) \right] \right], \tag{3.126}$$

$$S_H = \int_0^\beta du \left\{ \sum_{i,j \neq 0, \sigma} c_{i\sigma}^\dagger(u) \left[(\partial_u - \mu)\delta_{i,j} + t_{ij} \right] c_{j\sigma}(u) + U \sum_{i \neq 0} n_{i\uparrow}(u) n_{i\downarrow}(u) \right\}. \tag{3.127}$$

Here, S_0 denotes the contribution from the 0th site, S_{0H} from the coupling between the 0th site and the others, and S_H from the other sites. Using this decomposition, the partition function can be rewritten as

$$Z = Z_H \int \mathcal{D}c_{0,\sigma}^\dagger \mathcal{D}c_{0,\sigma} \exp(-S_0) \langle \exp(-S_{0H}) \rangle_H \tag{3.128}$$

$$Z_H = \int \prod_{i \neq 0, \sigma} \mathcal{D}c_{i,\sigma}^\dagger \mathcal{D}c_{i,\sigma} \exp(-S_H), \tag{3.129}$$

where $\langle A \rangle_H$ is the statistical average of an operator A using the action S_H, namely,

$$\langle A \rangle_H = \frac{1}{Z_H} \int \prod_{i \neq 0, \sigma} \mathcal{D}c_{i,\sigma}^\dagger \mathcal{D}c_{i,\sigma} A \exp(-S_H). \tag{3.130}$$

In order to integrate out the contribution from sites except for the 0th site, let us expand $\langle \exp(-S_{0H}) \rangle_H$ to which all the sites contribute in the powers of S_{0H}:

$$\langle \exp(-S_{0H}) \rangle_H = 1 - \langle S_{0H} \rangle_H + \frac{1}{2} \langle (S_{0H})^2 \rangle_H \cdots. \tag{3.131}$$

Due to the conservation of the total particle number, the first-order term equals 0. The second-order term can be written down as

$$\frac{1}{2} \langle (S_{0H})^2 \rangle_H = \int_0^\beta du_1 du_2 \sum_\sigma c_{0\sigma}^\dagger(u_1) \left[\sum_{i,j} t_{i0} t_{0j} \left\langle T_u c_{i\sigma}(u_1) c_{j\sigma}^\dagger(u_2) \right\rangle_H \right] c_{0\sigma}(u_2). \tag{3.132}$$

3.4 Dynamical Mean-Field Theory and Its Combination with FLEX

In order to simplify the problem, let us consider the d(spatial dimension)-dependence of the Green's function in the limit of large d. It is important to note that the limit of large d necessitates rescaled hopping integrals $t_{ij}/\sqrt{d}$ to assure the band width to remain finite in $d \to \infty$. Whether a perturbation term is relevant in $d \to \infty$ is determined by a balance between this rescaling factor $1/\sqrt{d}$ and the factor d coming from the summation. Since at least two hopping processes are necessary to hop between ith and jth sites, the single-particle Green's function gives a factor of $1/d$. Therefore the second-order term can not be neglected in the limit $d \to \infty$. Similarly for the nth order term, paying an attention to a factor d^{1-n} arising from the n-point Green's function, we can show that only the terms which can be decomposed into a product of the single-particle Green's function contribute to the expansion:

$$\langle \exp(-S_{0H}) \rangle_H \simeq \exp\left\{ -\int_0^\beta du_1 du_2 \sum_\sigma c_{0\sigma}^\dagger(u_1) \left[\sum_{i,j} t_{i0} t_{0j} G_{ij\sigma}^{(0)}(u_1 - u_2) \right] c_{0\sigma}(u_2) \right\} \tag{3.133}$$

Here we have introduced the Green's function in the system with the 0th site hollowed out as

$$G_{ij\sigma}^{(0)}(u_1 - u_2) = -\left\langle T_u c_{i\sigma}(u_1) c_{j\sigma}^\dagger(u_2) \right\rangle_H. \tag{3.134}$$

Summarizing the above discussion, the effective action for the 0th site can be expressed as

$$S_{\text{eff}} = -\int_0^\beta du_1 du_2 \sum_\sigma c_{0\sigma}^\dagger(u_1) \mathcal{G}_\sigma^{-1}(u_1 - u_2) c_{0\sigma}(u_2) + U \int_0^\beta du\, n_{0\uparrow}(u) n_{0\downarrow}(u), \tag{3.135}$$

by defining the effective-field Green's function as

$$\mathcal{G}_\sigma^{-1}(u) = -\partial_u + \mu - \sum_{ij} t_{i0} t_{0j} G_{ij\sigma}^{(0)}(u). \tag{3.136}$$

In the Matsubara representation, this becomes

$$\mathcal{G}_\sigma^{-1}(i\omega_n) = i\omega_n + \mu - \sum_{ij} t_{i0} t_{0j} G_{ij\sigma}^{(0)}(i\omega_n). \tag{3.137}$$

By comparing Eq. (3.135) to Eq. (3.120), we can clearly see the correspondence between the Hubbard model and the Anderson impurity model. In order to determine the Green's function $G_{ij\sigma}$, one has to calculate the statistical average in the system with the 0th site hollowed out. Note that the relationship between $G_{ij\sigma}^{(0)}$ and the Green's function in the original lattice system $G_{ij\sigma}$ is given by

$$G_{ij\sigma}^{(0)}(i\omega_n) = G_{ij\sigma}(i\omega_n) - \frac{\sum_i G_{i0\sigma}(i\omega_n) G_{0j\sigma}(i\omega_n)}{G_{00\sigma}(i\omega_n)}. \tag{3.138}$$

This equation can be proven by using the expansion of Green's functions in the hopping matrix elements [12]. Therefore we have

$$\sum_{ij} t_{i0} t_{0j} G_{ij\sigma}^{(0)}(i\omega_n) = \sum_{ij} t_{i0} t_{0j} G_{ij\sigma}(i\omega_n) - \frac{\left(\sum_i t_{0i} G_{i0\sigma}(i\omega_n)\right)^2}{G_{00\sigma}(i\omega_n)}. \tag{3.139}$$

When the self-energy is approximated as momentum-independent $\Sigma(k) = \Sigma(i\omega_n)$, the lattice Green's function can be written as

$$G_\sigma(k) = \frac{1}{i\omega_n + \mu - \epsilon(k) - \Sigma_\sigma(i\omega_n)}. \tag{3.140}$$

Using the Fourier series expansion,

$$G_{ij\sigma}(i\omega_n) = \sum_k \frac{e^{ik \cdot (r_i - r_j)}}{i\omega_n + \mu - \epsilon(k) - \Sigma_\sigma(i\omega_n)}, \tag{3.141}$$

where r_i is the position of the ith site. Therefore we have

$$\sum_{ij} t_{i0} t_{0j} G_{ij\sigma}^{(0)}(i\omega_n) = \sum_k G_\sigma(k) (\epsilon(k))^2 - \frac{\left(\sum_k G_\sigma(k) \epsilon(k)\right)^2}{G_{00\sigma}(i\omega_n)}. \tag{3.142}$$

Here the summation over k implies the normalization factor of N^{-1} (N is the number of k points). By definition, the local Green's function at the 0th site satisfies

$$G_{00\sigma}(i\omega_n) = \sum_k G_\sigma(k). \tag{3.143}$$

Therefore we can obtain the following equations by replacing the summation over k with the integral over the frequency:

$$\frac{1}{N} \sum_k G_\sigma(k) \epsilon(k) = i\omega_n + \mu - \epsilon(k) - \Sigma_\sigma(i\omega_n) - 1 \tag{3.144}$$

$$\frac{1}{N} \sum_k G_\sigma(k) (\epsilon(k))^2 = -(i\omega_n + \mu - \epsilon(k) - \Sigma_\sigma(i\omega_n))$$

$$+ (i\omega_n + \mu - \epsilon(k) - \Sigma_\sigma(i\omega_n))^2 G_{00\sigma} \tag{3.145}$$

Here the normalization factor of N^{-1} is explicitly written. Using these equations, we have

3.4 Dynamical Mean-Field Theory and Its Combination with FLEX

$$\mathcal{G}_\sigma^{-1}(k) = \left[\frac{1}{N}\sum_k G_\sigma(k)\right]^{-1} + \Sigma_\sigma(i\omega_n). \quad (3.146)$$

From the above discussion, we can construct the DMFT self-consistent scheme as follows:

1. Start from an initial guess for the self-energy $\Sigma_\sigma(i\omega_n)$.
2. Calculate the local Green's function:

$$G_{00\sigma}(i\omega_n) = \frac{1}{N}\sum_k G_\sigma(k) \quad (3.147)$$

3. Calculate the effective field Green's function by Eq. (3.146).
4. Solve the impurity problem by treating $\mathcal{G}_\sigma(i\omega_n)$ as the non-interacting Green's function and update the self-energy.
5. Go back to (2) (and iterate until convergence).

DMFT is a typical non-perturbative method and can describe the Mott transition, which is a metal-insulator transition due to the strong local correlation. Another, and quite important advantage is that DMFT can be formulated from the Luttinger–Ward functional in which diagrams consisting of the purely local Green's function are taken into account [25]. Namely, the DMFT is a conserving approximation, which is quite useful in constructing a generalization to systematically improve the approximation.

3.4.3 Impurity Solvers

DMFT maps a lattice problem onto a single impurity problem. The self-energy in this impurity problem can be obtained much easier than the original one. There are many methods for the impurity problem (impurity solvers). Here we introduce two methods: the modified iterated perturbation theory (mIPT), which is based on the weak-coupling perturbation theory, and the continuous-time quantum Monte Carlo (CT-QMC), which is numerically exact.

Modified Iterated Perturbation Theory

The iterated perturbation theory (IPT) is a solver for the impurity model using the second-order perturbation theory [11, 26–28]. In the original IPT, the first-order term of the self-energy $\Sigma^{(1)}$ is given by

$$\Sigma^{(1)}(i\omega_n) = U\frac{n}{2}. \quad (3.148)$$

Here n is the number of electrons per unit cell which is called the band filling. The second order-term can be easily derived by a simple perturbative calculation with respect to U as

$$\Sigma^{(2)}(u) = U^2 \mathcal{G}(u)\mathcal{G}(-u)\mathcal{G}(u). \tag{3.149}$$

This scheme works surprisingly well in the electron-hole symmetric case since the self-energy within this approximation reproduces the asymptotic behavior of the exact self-energy in the atomic and high-frequency limit in such a case. However, in electron-hole *asymmetric* cases, this relation is broken down and hence IPT does not give good results.

In the modified IPT, the self-energy is parametrized by two parameters to resolve this issue:

$$\Sigma(i\omega_n) = U\frac{n}{2} + \frac{A\Sigma^{(2)}(i\omega_n)}{1 - B\Sigma^{(2)}(i\omega_n)}, \tag{3.150}$$

where A and B are introduced so as to reproduce the exact self-energy in the atomic and high-frequency limit, and can be written as

$$A = \frac{n(1-n)}{n_0(1-n_0)}, \quad B = \frac{(1-n)U + \mu_0 - \mu}{n_0(1-n_0)U^2} \tag{3.151}$$

$$n = \sum_n G(i\omega_n) + \frac{1}{2}, \quad n_0 = \sum_n \mathcal{G}(i\omega_n) + \frac{1}{2}. \tag{3.152}$$

One can easily show that this self-energy is reduced to the original IPT in the electron-hole symmetric case. In electron-hole asymmetric cases, a constraint is necessary to determine the parameters. There are several versions such as requiring the Luttinger sum rule, introducing constraint on the occupancy as $n = n_0$ by hand, or treating the double occupancy as an input. Since the parametrization is valid only for single-band systems, one has to adopt other solvers in order to treat multi-band systems.

Continuous-Time Quantum Monte Carlo

The continuous-time quantum Monte Carlo is a family of several algorithms for solving the impurity model in a numerically exact way by expanding the partition function with the aid of the Monte Carlo integration [29–31]. Broadly speaking, there are two common choices of the expansion for the impurity problem with general local interactions: the continuous-time interaction expansion (CT-INT) and the continuous-time hybridization expansion (CT-HYB). The partition function is expanded with respect to the local interaction term in the former choice. The latter is based on the expansion in terms of the hybridization between the conduction electron (bath) and the impurity. Here we briefly describe CT-HYB, which is used in the present study, referencing Ref. [32].

3.4 Dynamical Mean-Field Theory and Its Combination with FLEX

Let us consider a multi-orbital impurity Hamiltonian with a general on-site interaction term:

$$\mathcal{H} = \mathcal{H}_{\text{loc}} + \mathcal{H}_{\text{bath}} + \mathcal{H}_{\text{hyb}}, \tag{3.153}$$

$$\mathcal{H}_{\text{loc}} = \sum_{ab} t_{ab} c_a^\dagger c_b + \sum_{abcd} U_{abcd} c_a^\dagger c_b^\dagger c_c c_d, \tag{3.154}$$

$$\mathcal{H}_{\text{bath}} = \sum_{\alpha} \epsilon_\alpha d_\alpha^\dagger d_\alpha, \tag{3.155}$$

$$\mathcal{H}_{\text{hyb}} = \sum_{\alpha b} \left[V_{\alpha b} d_\alpha^\dagger c_b + \text{h.c.} \right]. \tag{3.156}$$

The indices a and b mean the internal degrees of freedom (i.e. spin and orbital) of the impurity, whereas α those (i.e. spin, orbital, and momentum) of the bath. t_{ab} contains the chemical potential for simplicity. The effective action can be derived by tracing out the bath degrees of freedom:

$$S_{\text{eff}} = \int_0^\beta du\, \mathcal{H}_{\text{loc}}(u) + \int_0^\beta du_1 du_2 \sum_{ab} c_a^\dagger(u_1) \Delta_{ab}(u_1 - u_2) c_b(u_2), \tag{3.157}$$

$$\Delta_{ab} = \sum_{\alpha} \frac{V_{a\alpha}^* V_{\alpha b}}{i\omega_n - \epsilon_\alpha} \tag{3.158}$$

The expansion of the grand partition function of the impurity model in powers of the hybridization can be expressed as

$$\Xi \equiv \langle \exp(-\beta \mathcal{H}) \rangle$$

$$\propto \sum_{n=0}^{\infty} \frac{1}{n!^2} \sum_{\{a_i\},\{a_i'\}} \int_0^\beta du_1 du_1' \ldots \int_0^\beta du_n du_n'$$

$$\times \text{Tr} \left\{ e^{u\mathcal{H}_{\text{loc}}} T_u c_{a_1}(u_1) c_{a_1'}^\dagger(u_1') \ldots c_{a_n}(u_n) c_{a_n'}^\dagger(u_n') \right\}_{\text{loc}} \det M^{-1}. \tag{3.159}$$

Here we explicitly show the discretized imaginary time u and $A(u) = e^{u\mathcal{H}_{\text{loc}}} A e^{-u\mathcal{H}_{\text{loc}}}$. M is defined by the hybridization function

$$\left(M^{-1}\right)_{ij} = \Delta_{a_i' a_j}(u_i' - u_j). \tag{3.160}$$

Equation (3.159) can be also written as

$$\Xi \propto \sum_{n=0}^{\infty} \sum_{\{a_i\},\{a_i'\}} \int_0^{u_2} du_1 \int_0^{u_2'} du_1' \ldots \int_0^\beta du_n \int_0^\beta du_n'$$

$$\times \text{Tr} \left\{ e^{-(\beta - \tilde{u}_{2n})\mathcal{H}_{\text{loc}}} O_{2n} \ldots e^{-(\tilde{u}_2 - \tilde{u}_1)\mathcal{H}_{\text{loc}}} O_1 e^{-\tilde{u}_1 \mathcal{H}_{\text{loc}}} \right\}_{\text{loc}} (-1)^P \det M^{-1}. \tag{3.161}$$

Here $0 \leq u_1 \ldots \leq u_n$, $0 \leq u'_1 \ldots \leq u'_n$, $\{O_1, \ldots, O_{2n}\}$ is a time ordered set of cs and $c^\dagger$s, and $\{\tilde{u}_1, \ldots, \tilde{u}_{2n}\}$ a time ordered set of imaginary time variables. P_{trace} is the permutation of the time ordering. In the Monte Carlo simulation, an importance sampling of Z using the configurations of u and a is performed.

As for the single-particle Green's function defined as

$$G_{ab}(u_1 - u_2) = -\frac{\text{Tr}\left\{T_u e^{-S_{\text{eff}}} c_a(u_1) c_b^\dagger(u_2)\right\}}{Z}, \tag{3.162}$$

similarly to the grand partition function, the expansion of the numerator can be expressed as

$$\text{Tr}\left\{T_u e^{-S_{\text{eff}}} c_a(u) c_b^\dagger(u')\right\} \propto \sum_{n=0}^{\infty} \sum_{\{a_i\},\{a'_i\}} \int_0^{u_2} du_1 \int_0^{u'_2} du'_1 \ldots \int_0^{\beta} du_n \int_0^{\beta} du'_n$$
$$\times \text{Tr}\left\{e^{-(\beta-\tilde{u}_{2n+2})\mathcal{H}_{\text{loc}}} O_{2n+2} \ldots e^{-(\tilde{u}_2-\tilde{u}_1)\mathcal{H}_{\text{loc}}} O_1 e^{-\tilde{u}_1 \mathcal{H}_{\text{loc}}}\right\}_{\text{loc}}$$
$$(-1)^P \det \boldsymbol{M}^{-1}. \tag{3.163}$$

In practice, instead of the representation in imaginary time, it is useful to expand the Green's function in the Legendre polynomials defined on the interval $[0, \beta)$ as proposed in [33]:

$$G_{ab}(u) = \sum_{l \geq 0}^{n_{\text{Leg}}} \frac{\sqrt{2l+1}}{\beta} P_l\left(\frac{2u}{\beta} - 1\right) G_l^{ab},$$
$$G_l^{ab} = \sqrt{2l+1} \int_0^{\beta} du\, P_l\left(\frac{2u}{\beta} - 1\right) G_{ab}(u), \tag{3.164}$$

where P_l is the lth Legendre polynomial and n_{Leg} is the cutoff for the Legendre polynomial expansion, whose typical values are $n_{\text{Leg}} = 50$–100. In the Monte Carlo simulation, an importance sampling of G_l^{ab} is performed.

In the present study, we solve the impurity problem by the CT-HYB algorithm described above with codes based on the ALPSCore libraries and ALPSCore/CT-HYB [32, 34, 35].

3.4.4 FLEX+DMFT Method

The FLEX approximation is based on the perturbation theory with respect to the interaction term. And importantly, it is guaranteed to be a conserving approximation since one can formulate it by collecting the bubble- and ladder- type diagrams in the Luttinger–Ward functional. Thanks to this, one may treat non-local fluctuation effects

3.4 Dynamical Mean-Field Theory and Its Combination with FLEX

such as magnetic/charge fluctuations in a self-consistent manner. However, the vertex corrections, which coming from the higher-order correction terms, are neglected in the FLEX approximation. Due to this shortcoming, it can not sufficiently capture the strong correlation effects, exemplified by the strong quasi-particle renormalization seen in the vicinity of the Mott transition.

On the other hand, DMFT, a kind of mean-field approximation which neglects spatial fluctuations as the name suggests, can treat dynamical fluctuation accurately. While this approximation becomes exact in the limit of large spatial dimension, it is not justified in low dimensional systems since strong spatial fluctuations play a key role. Another important feature is that this approximation can be formulated via the Luttinger–Ward functional, namely guaranteed as conserving.

Namely, if we classify the correlation effects into the local and non-local ones, whereas the former can be well captured within the DMFT, the latter can be well captured within the FLEX approximation.

Along this line, the FLEX+DMFT method is a promising approach, at least in the intermediate coupling region, to incorporate both local and non-local electron correlations in a self-consistent manner. This method is a conserving scheme in the Baym–Kadanoff sense. This method was first formulated for a single band system and was applied to the single band Hubbard model on a square lattice in Refs. [21–23]. It can capture the so-called pseudogap behavior in the single-particle spectrum in the single-band Hubbard model on a square lattice [22], which is considered as a typical effect of the Mott physics, and is consistent with experiments on high-T_c cuprates [36, 37] and the dual fermion method [38].

Since our aim is to study superconductivity in multiband systems, here we extend the FLEX+DMFT formulation so that it can be applied to multiband systems. Let us consider the multi-band Hubbard model, in which the unit cell comprises multiple atomic sites but there is no orbital degrees of freedom in each site. Namely, the Hamiltonian reads

$$\mathcal{H} = \sum_{ij\sigma}\sum_{\alpha\beta} t_{ij}^{\alpha\beta} c_{i\sigma}^{\alpha\dagger} c_{j\sigma}^{\beta} + \sum_i \sum_\alpha U_\alpha n_{i\uparrow}^\alpha n_{i\downarrow}^\alpha. \tag{3.165}$$

Formal derivation is completely the same as in the single-band case. Namely we employ an approximated form of the Luttinger–Ward functional as

$$\Phi_{\text{FLEX+DMFT}}[\hat{G}] = \Phi_{\text{FLEX}}[\hat{G}] - \Phi_{\text{FLEX}}^{\text{local}}[\hat{G}_{\text{loc}}] + \Phi_{\text{DMFT}}[\hat{G}_{\text{loc}}], \tag{3.166}$$

where $\Phi_{\text{FLEX(DMFT)}}[\hat{G}]$ is the Luttinger–Ward functional for the FLEX approximation (DMFT). $\Phi_{\text{FLEX}}^{\text{local}}[\hat{G}_{\text{loc}}]$ is the Luttnger–Ward functional collecting bubble- and ladder-type diagrams comprising the local Green's function $\hat{G}_{\text{loc}}$, which subtracts the double counting term in the FLEX and DMFT functional. $\hat{A}$ emphasizes that a quantity A is in the matrix representation with the site indices. Taking the functional derivative with respect to $\hat{G}$, we obtain the self-energy within the approximation:

$$\hat{\Sigma}_{\text{FLEX+DMFT}}[\hat{G}] = \hat{\Sigma}_{\text{FLEX}}[\hat{G}] - \hat{\Sigma}_{\text{FLEX}}^{\text{local}}[\hat{G}_{\text{loc}}] + \hat{\Sigma}_{\text{DMFT}}[\hat{G}_{\text{loc}}] \qquad (3.167)$$

Our implementation of this scheme for the numerical calculation can be summarized as follows, which is slightly different from that in Ref. [22], but we have confirmed that it gives the same results:

1. Start from an initial guess for the self-energy $\hat{\Sigma}(k)$ (usually set to 0). Here we abbreviate $\hat{\Sigma}_{\text{FLEX+DMFT}}$ as $\hat{\Sigma}(k)$.
2. Calculate the self-energy by

$$\hat{\Sigma}(k) = \hat{\Sigma}_{\text{FLEX}}(k) - \hat{\Sigma}_{\text{FLEX}}^{\text{local}}(i\omega_n) + \hat{\Sigma}_{\text{DMFT}}(i\omega_n). \qquad (3.168)$$

3. Obtain the Green's function by solving the Dyson equation:

$$\hat{G}(k) = \left[(i\omega_n + \mu)\hat{I} - \hat{\mathcal{H}}(k) - \hat{\Sigma}(k)\right]^{-1}. \qquad (3.169)$$

At this step, one has to determine the chemical potential so that the number of electrons is unchanged during the calculation. To reduce numerical errors due to the slow-decaying asymptotic behavior

$$G_{\alpha\alpha}(i\omega_n) \sim \frac{1}{i\omega_n} \qquad (3.170)$$

in the high-frequency region, we recast the sum rule for the Green's function as

$$n = 2\frac{k_{\text{B}}T}{N}\sum_{k\alpha} G_{\alpha\alpha}(k) + 1$$
$$= 2\frac{k_{\text{B}}T}{N}\sum_{k\alpha} \left[G_{\alpha\alpha}(k) - G_{\alpha\alpha}^{(0)}(k)\right] + \frac{2}{N}\sum_{kn} f(\epsilon_n(k)) \qquad (3.171)$$

Here f is the Fermi–Dirac distribution function, and $\epsilon_n(k)$ the nth eigenvalue of $\hat{H}(k)$. The factor 2 implies the spin degeneracy. To estimate the chemical potential, we employ the bisection method whose interval is determined with the linear search method.

4. Calculate the local Green's function by

$$\hat{G}_{\text{loc}} = \frac{1}{N}\sum_k \hat{G}(k). \qquad (3.172)$$

And obtain the hybridization function $\hat{\Delta}$ used as an input for the impurity problem. Specifically for CT-HYB, since the hybridization function $\hat{\Delta}$ in the imaginary time representation is used as an input in the ALPSCore/CT-HYB code, the Fourier transformation from the Matsubara representation is necessary. We subtract $\mathcal{O}(i\omega_n^{-1})$ component determined with the non-linear least square method in

3.4 Dynamical Mean-Field Theory and Its Combination with FLEX

the high-frequency region, and add the corresponding term calculated analytically after the Fourier transformation.
5. Calculate the DMFT part of the self-energy with an impurity solver (e.g. CT-HYB or mIPT).
6. Calculate the FLEX part of the self-energy by

$$\Sigma_{lm}(k) = \frac{1}{2}\frac{k_B T}{N} \sum_q \Gamma_{llmm}(q) G_{lm}(k-q) \quad (3.173)$$

$$\hat{\Gamma}(q) = \hat{U} \left\{ \frac{3}{2} \left[\hat{I} - \hat{\chi}^{\text{irr}}(q)\hat{U}\right]^{-1} \hat{\chi}^{\text{irr}}(q) \right.$$
$$\left. + \frac{1}{2} \left[\hat{I} + \hat{\chi}^{\text{irr}}(q)\hat{U}\right]^{-1} \hat{\chi}^{\text{irr}}(q) + \hat{\chi}^{\text{irr}}(q) \right\} \hat{U} + \hat{U}. \quad (3.174)$$

Here $\hat{U} \equiv U\hat{I}$ is the interaction vertex. And calculate $\hat{\Sigma}^{\text{local}}_{\text{FLEX}}$ similarly. Since the maximum value α_S of the eigenvalues of $\hat{\chi}^{\text{irr}}(q)\hat{U}$ often becomes $\alpha_S > 1$ at an early stage of the self-consistent calculation in the presence of strong spatial fluctuations, Γ exhibits singular behavior. Since this leads to unphysical hybridization functions during the calculation, and the calculation becomes unstable specifically in the case of the CT-HYB solver. In order to avoid this numerical instability, we rescale $\hat{\chi}^{\text{irr}}(q)$ as

$$\hat{\chi}^{\text{irr}}(q) = \frac{1-\eta}{\alpha_S} \hat{\chi}^{\text{irr}}(q) \quad (3.175)$$

if $\alpha_S > 1 - \eta$ in each iteration. Here η is a small constant, say $\eta = 10^{-3}$. Note that this rescaling is done only to avoid the numerical instability. If the trial self-energy is sufficiently close to the actual solution, it is no longer necessary. Therefore the converged solution is not affected by this procedure. We have confirmed that the numerical stability is improved by this method, and furthermore the convergence speed becomes (typically 1.5–2 times) faster.
7. Go back to 2 and iterate until convergence.

As for calculations for superconductivity, since the Luttinger–Ward formalism can be extended to the functional containing anomalous quantities, the anomalous self-energy Δ can be derived by the functional derivative with respect to F:

$$\Delta_{\text{FLEX+DMFT}} = \Delta_{\text{FLEX}} + \Delta_{\text{DMFT}} - \Delta_{\text{FLEX}}^{\text{local}}. \quad (3.176)$$

If we consider the non-local pairing whose integral over the momentum becomes zero (e.g. $d_{x^2-y^2}$-wave in cuprates), we can omit the local correction term. If we omit the local correction, the linearized Eliashberg equation is the same as that of the FLEX approximation:

$$\lambda \Delta_{ll'}(k) = -\frac{T}{N} \sum_{k'm_i} \Gamma^s_{lm_1 m_4 l'}(k-k') G_{m_1 m_2}(k') \Delta_{m_2 m_3}(k') G_{m_4 m_3}(-k'), \quad (3.177)$$

$$\hat{\Gamma}^s(q) = \hat{U} \left\{ \frac{3}{2} \left[\hat{I} - \hat{\chi}^{\text{irr}}(q)\hat{U} \right]^{-1} \hat{\chi}^{\text{irr}}(q) - \frac{1}{2} \left[\hat{I} + \hat{\chi}^{\text{irr}}(q)\hat{U} \right]^{-1} \hat{\chi}^{\text{irr}}(q) \right\} \hat{U} + \hat{U}. \quad (3.178)$$

3.5 Density Functional Theory

First-principles band calculation is a promising way to study electronic properties of actual materials. In this section, we will review some basics of the density functional theory (DFT) and the calculation methods based on it, referencing Refs. [39–41].

3.5.1 Born-Oppenheimer Approximation

Let us consider the non-relativistic Hamiltonian for a crystalline solid comprising nuclei and electrons,

$$\begin{aligned} \mathcal{H} &= T_e + T_N + U \\ U &= U_{ee} + U_{NN} + U_{eN}, \end{aligned} \quad (3.179)$$

where $T_{e(N)}$ is the kinetic energy of electrons (nuclei), $U_{ee(NN)}$ is the Coulomb interaction between electrons (nuclei), and U_{eN} is the electron–nucleus Coulomb interaction:

$$T_e = -\sum_i \frac{\nabla_i^2}{2}, \quad T_N = -\sum_j \frac{\nabla_j^2}{2M_j}, \quad U_{ee} = \frac{1}{2} \sum_{i \neq i'} \frac{1}{|\mathbf{r}_i - \mathbf{r}_{i'}|},$$

$$U_{NN} = \frac{1}{2} \sum_{j \neq j'} \frac{Z_j Z_{j'}}{|\mathbf{R}_j - \mathbf{R}_{j'}|}, \quad U_{eN} = -\frac{1}{2} \sum_{ij} \frac{Z_j}{|\mathbf{r}_i - \mathbf{R}_j|}, \text{ in the Hartree unit.}$$

$$(3.180)$$

Here i (j) labels electrons (nuclei) and M is the nucleus mass, as well as $\mathbf{r}$ ($\mathbf{p}$) denotes the position (momentum) of an electron and $\mathbf{R}$ ($\mathbf{P}$) that of a nucleus.

If the Schrödinger equation for this Hamiltonian can be solved, one can obtain all of the information in this system. However, in reality, this is impossible since the Hamiltonian is for a many-body system containing both electron and nucleus degrees of freedom. To reduce the complexity of the problem, here we apply an adiabatic approximation which is called the Born–Oppenheimer approximation. Namely, since nuclei is much more massive than electrons, we can approximate that the position of the nuclei as being fixed with respect to the electron motion. Thus, within this

3.5 Density Functional Theory

approximation, if we are interested in the electronic structure, the electron–nucleus interaction U_{eN} can be treated as the external potential V_{ext} in the electronic Hamiltonian:

$$\mathcal{H} = T_e + U_{ee} + \sum_i V_{\text{ext}}(\mathbf{r}_i). \tag{3.181}$$

This can be expressed also in the second quantization representation as

$$\mathcal{H} = -\sum_\sigma \int \psi_\sigma^\dagger(\mathbf{r}) \frac{\nabla^2}{2} \psi_\sigma(\mathbf{r}) d\mathbf{r} + \sum_\sigma \int \psi_\sigma^\dagger(\mathbf{r}) V_{\text{ext}}(\mathbf{r}) \psi_\sigma(\mathbf{r}) d\mathbf{r}$$
$$+ \frac{1}{2} \sum_{\sigma\sigma'} \int d\mathbf{r} \int d\mathbf{r}' \psi_\sigma^\dagger(\mathbf{r}) \psi_{\sigma'}^\dagger(\mathbf{r}') \frac{1}{|\mathbf{r}-\mathbf{r}'|} \psi_{\sigma'}(\mathbf{r}') \psi_\sigma(\mathbf{r}). \tag{3.182}$$

Note that the expression within the Born–Oppenheimer approximation contains an important feature on the electronic structure in actual materials: since the first and third terms are purely electronic, the material dependence of the electronic structure arises only from the external potential V_{ext}. This feature plays an important role in the construction of DFT.

3.5.2 Hohenberg-Kohn Theorem

Using the Born–Oppenheimer approximation, the Hamiltonian is separated into the electronic and the nuclear parts. However, actual calculations remain difficult since we have to treat the many-body wave function as is. Here we introduce the Hohenberg–Kohn theorem [42], which allows us to find the ground state properties of a many-body system *without* dealing directly with the many-body wave function.

For convenience, we rewrite the Hamiltonian introducing the parameter λ ($0 \leq \lambda \leq 1$) controlling the intensity of the interaction term U_{ee},

$$\mathcal{H}_e(\lambda) = T_e + V_{\text{ext}} + \lambda U_{ee}. \tag{3.183}$$

Let us consider a trial state $|\Phi_{\text{tri}}\rangle$ for N-body system with the normalization condition,

$$\langle \Phi_{\text{tri}} | \sum_\sigma \int d\mathbf{r} \psi_\sigma^\dagger(\mathbf{r}) \psi_\sigma(\mathbf{r}) |\Phi_{\text{tri}}\rangle \equiv \int d\mathbf{r} \rho(\mathbf{r}) = N, \tag{3.184}$$

where $\rho(\mathbf{r})$ is the electron density. Using this trial state, the expectation value of the total energy E can be written as

$$E = \langle \Phi_{\text{tri}} | T_e + \lambda U_{ee} | \Phi_{\text{tri}} \rangle + \langle \Phi_{\text{tri}} | V_{\text{ext}} | \Phi_{\text{tri}} \rangle. \tag{3.185}$$

One can easily show that this total energy satisfies the so-called Schödinger–Ritz variational principle,

$$E \geq E_0, \tag{3.186}$$

where $E_0(\lambda)$ is the ground state energy and the equality holds when $|\Phi\rangle$ is the ground state.

With this, we describe the Hohenberg–Kohn theorem:

Theorem 3.1 *An external potential V_{ext} is a unique functional of the electron density $\rho_\lambda(r)$ in the system characterized by $\mathcal{H}_e(\lambda)$, apart from a trivial constant.*

Proof Assume that there exists V_{ext} for any $\rho(r)$ (V-representability) and the ground state is not degenerated (non-degenerate ground state). Note that though it is known that this theorem can be generalized to the case of degenerate ground state, we employ the assumption of non-degenerate ground state for simplicity.

Assume that there exist two external potentials $V_e(r)$ and $V'_e(r)$ which differ by more than a constant and corresponds to the same ground state density $\rho_\lambda(r)$. Let $|\Phi_0^{(\prime)}\rangle$ and $E_0^{(\prime)}$ be the ground state and the energy of the Hamiltonian characterized by $V_e^{(\prime)}(r)$. If $|\Phi_0\rangle = |\Phi'_0\rangle$,

$$(T_e + \lambda U_{ee})|\Phi_0\rangle + \sum_\sigma \int dr\, \phi_\sigma^\dagger(r) V_{\text{ext}}(r) \phi_\sigma(r) |\Phi_0\rangle = E_0 |\Phi_0\rangle, \tag{3.187}$$

$$(T_e + \lambda U_{ee})|\Phi_0\rangle + \sum_\sigma \int dr\, \phi_\sigma^\dagger(r) V'_{\text{ext}}(r) \phi_\sigma(r) |\Phi_0\rangle = E'_0 |\Phi_0\rangle, \tag{3.188}$$

hold. Thus we have

$$V_{\text{ext}}(r) - V'_{\text{ext}}(r) = E_0 - E'_0. \tag{3.189}$$

This is in contradiction with the assumption that $V_e(r)$ and $V'_e(r)$ differ by more than a constant. Therefore, two states $|\Phi_0\rangle$ and $|\Phi'_0\rangle$ are distinct.

Using the Schrödinger-Ritz variational principle,

$$\begin{aligned} E_0 &= \langle \Phi_0 | T_e + \lambda U_{ee} | \Phi_0 \rangle + \int dr\, V_{\text{ext}} \rho_\lambda(r) \\ &< \langle \Phi'_0 | T_e + \lambda U_{ee} | \Phi'_0 \rangle + \int dr\, V_{\text{ext}} \rho_\lambda(r) \\ &= E'_0 + \int dr\, \left[V_{\text{ext}}(r) - V'_{\text{ext}}(r) \right] \rho_\lambda(r). \end{aligned} \tag{3.190}$$

Clearly we also have,

$$E_0 < E'_0 + \int dr\, \left[V'_{\text{ext}}(r) - V_{\text{ext}}(r) \right] \rho_\lambda(r). \tag{3.191}$$

3.5 Density Functional Theory

Adding these inequality,

$$E_0 + E_0' < E_0 + E_0'. \tag{3.192}$$

This is clearly a contradiction. Therefore the theorem is proven by *reductio ad absurdum*.

Since the ground state $|\Phi\rangle$ is a functional of the electron density, the kinetic and interaction energies are also functionals of the electron density. Given this, let us define a "density functional":

$$\begin{aligned} E_{V_{\text{ext}}}[\rho(r)] &= \langle \Phi | T_e + \lambda U_{ee} | \Phi \rangle + \langle \Phi | V_{\text{ext}} | \Phi \rangle \\ &= F_{\text{HK}}[\rho(r)] + \langle \Phi | V_{\text{ext}} | \Phi \rangle . \end{aligned} \tag{3.193}$$

Here, since the first term $F_{\text{HK}}[\rho(r)]$ is independent of the external field, it is called a universal functional. This functional $E_{V_{\text{ext}}}[\rho(r)]$ has the variational character as can be seen from the second Hohenberg–Kohn theorem, which we will describe here:

Theorem 3.2 *The exact ground state is the global minimum of the density functional $E_0[\rho(r)]$.*

Proof The proof can be done in a similar way to theorem 1. Let us write an external potential corresponding to a certain density $\rho_\lambda(r)$ as $V_{\text{ext}}(r; [\rho_\lambda(r)]) \equiv V'_{\text{ext}}(r)$. Also a state for density $\rho_\lambda(r)$, $|\Phi_0(\lambda; [\rho_\lambda(r)])\rangle \equiv |\Phi_0'\rangle$.

If $V'_{\text{ext}}(r) = V_{\text{ext}}(r)$, $|\Phi_0'\rangle$ is the true ground state $|\Phi_0\rangle$. Thus, in this case, $E_0(\lambda; [\rho_\lambda(r)]) = E_0(\lambda)$ holds.

Since $|\Phi_0'\rangle$ and $|\Phi_0\rangle$ always differ if $V'_{\text{ext}}(r) \neq V_{\text{ext}}(r)$, we have $E_0(\lambda; [\rho_\lambda(r)]) > E_0(\lambda)$ from the Schrödinger-Ritz variational principle. Therefore $E_0[\rho(r)]$ is always larger than the true ground state energy E_0 except for the ground state.

It is important to note that the Hohenberg–Kohn theorem remarkably reduces the many-body problem of N electrons with $3N$ spatial coordinates into that with three spatial coordinates, through the use of density functionals.

3.5.3 Kohn-Sham Equation

Thanks to the Hohenberg–Kohn theorem, the many-body problem has been formulated as a variational problem with respect to an electron density. It is, however, still rather unrealistic to obtain the exact expression of the universal functional $F_{\text{HK}}[\rho(r)]$. In order to take advantage of the power of DFT without losing accuracy of the theoretical framework, we follow the scheme of Kohn and Sham [43] mapping the actual system onto a fictitious system of non-interacting "electrons".

Let us adiabatically increase λ from $\lambda = 0$ (non-interacting system) to $\lambda = 1$ (original system). The chemical potential μ_1 for $\lambda = 1$ can be written as,

$$\mu_1 = \frac{\delta E_{\lambda=1}[\rho(r)]}{\delta \rho(r)}$$
$$= \frac{\delta F_{\lambda=1}[\rho(r)]}{\delta \rho(r)} + V_{\text{ext}}(r)$$
$$= \frac{\delta T_s[\rho(r)]}{\delta \rho(r)} + \int dr' \frac{\rho(r')}{|r-r'|} + V_{\text{xc}}(r) + V_{\text{ext}}(r), \quad (3.194)$$

where $T_s[\rho(r)] = \delta F_{\lambda=0}[\rho(r)]$, and $V_{\text{xc}}(r)$ is called the exchange-correlation potential defined by,

$$V_{\text{xc}}(r) \equiv \frac{\delta E_{\text{xc}}[\rho(r)]}{\delta \rho(r)}. \quad (3.195)$$

Here we introduce the exchange-correlation functional $E_{\text{xc}}[\rho(r)]$ using the density functional for the interaction energy $E_{ee}[\rho(r)]$ and the Hartree energy (classical Coulomb interaction energy):

$$E_{\text{xc}}[\rho(r)] = T[\rho(r)] - T_s[\rho(r)] + E_{ee}[\rho(r)] - E_H[\rho(r)], \quad (3.196)$$
$$E_H[\rho(r)] = \int dr \int dr' \frac{\rho(r)\rho(r')}{|r-r'|}, \quad (3.197)$$

The chemical potential μ_0 for $\lambda = 0$ can be expressed as follows by introducing the external potential giving rise to the same density as that for $\lambda = 1$:

$$\mu_0 = \frac{\delta T_s[\rho(r)]}{\delta \rho(r)} + V_{\text{ext}}(r; \lambda = 0; [\rho(r)]). \quad (3.198)$$

Subtracting the Eq. (3.194) from the Eq. (3.198),

$$V_{\text{ext}}(r; \lambda = 0; [\rho(r)]) = \int dr' \frac{\rho(r')}{|r-r'|} + V_{\text{xc}}(r) + V_{\text{ext}}(r)$$
$$= V_{\text{KS}}(r) + \mu_0 - \mu_1, \quad (3.199)$$

Since $\mu_0 - \mu_1$ is a constant, $V_{\text{KS}}(r)$ and $V_{\text{ext}}(r; \lambda = 0; [\rho(r)])$ give the same density. In addition, we can choose the external potential for the non-interacting system $V_{\text{ext}}(r; \lambda = 0; [\rho(r)])$ as $V_{\text{KS}}(r)$.

As discussed above, one can obtain the ground state density and the chemical potential for the interacting many-body system by solving the single-particle Schrödinger(-like) equation under the external potential $V_{\text{KS}}(r)$:

$$\mathcal{H}_{\text{KS}} \phi_i(r) = \epsilon_i \phi_i(r) \quad (3.200)$$
$$\mathcal{H}_{\text{KS}} = -\frac{\nabla^2}{2} + \int dr' \frac{\rho(r')}{|r-r'|} + V_{\text{xc}}(r) + V_{\text{ext}}(r). \quad (3.201)$$

3.5 Density Functional Theory

This is called the Kohn–Sham equation, and the "single-electron" orbital $\phi_i(r)$ the Kohn–Sham orbital.

The Kohn–Sham orbitals are closely related to the electron density, but on the other hand, it is an important problem whether we can find a physical meaning of the eigenvalue of the Kohn–Sham equation. Here we briefly introduce the Janak's [44] theorem which plays an important role in this point. In the Kohn–Sham scheme, an electron density is given by,

$$\rho(r) = \sum_i^N n_i |\phi_i(r)|^2, \tag{3.202}$$

with the occupation number of the Kohn–Sham orbitals n_i satisfying $n_i = 1$ for $0 \le i \le N$ and $n_i = 1$ for $i > N$. Using fractional occupation numbers ($0 \le n_i \le 1$ for each i) introduced by Janak, the kinetic term is modified as

$$\tilde{T}_s[n(r)] = \sum_i n_i \langle \phi | -\frac{\nabla^2}{2} | \phi \rangle. \tag{3.203}$$

Introducing the chemical potential μ_1 as a Lagrange multiplier for the thermodynamic potential Ω and minimizing Ω:

$$\Omega[\{n_i\}; \rho(r)] = E_0[\{n_i\}; \rho(r)] - \mu_1 N. \tag{3.204}$$

With this, the energy functional satisfies,

$$\left.\frac{\partial E_0}{\partial n_i}\right|_{\rho(r)} = \epsilon_i, \tag{3.205}$$

in the ground state. Therefore ϵ_i can be interpreted as the energy of an electron in the ith Kohn–Sham orbital.

Since the eigenfunction of the Kohn–Sham equation is the eigenstate of the single-particle problem with a periodic potential of the crystal, it is represented by the Bloch's theorem. So ϵ_i and $\phi_i(r)$ are often treated as the eigenenergy and the wave function of am electron in the crystal, respectively.

The actual calculation solving the Kohn–Sham equation must be a self-consistent scheme as one can easily understand from the derivation. In Fig. 3.8, we show a flowchart of the procedure.

3.5.4 Exchange-Correlation Functional

Since the results so far are exact, one can obtain the ground state properties of many-body systems, provided that the explicit form of $E_{xc}[n]$ is known. However, it is

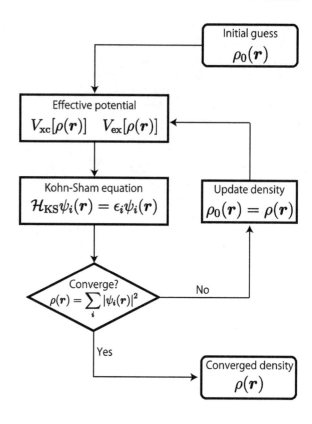

Fig. 3.8 A flowchart for the procedure to solve the Kohn–Sham equation

extremely difficult to know an exact form of $E_{xc}[\rho(r)]$ since $E_{xc}[\rho(r)]$ describes the most difficult part of the problem, namely many-body effect arising from the electron–electron interaction. Therefore an approximate form of $E_{xc}[\rho(r)]$ is required for an actual calculation. Here we introduce two simple approximations that work well: Local Density Approximation (LDA) and Generalized Gradient Approximation (GGA).

It is important to note that the exchange-correlation functional (Eq. 3.197) is defined separately from the Hartree term which describes the classical Coulomb interaction. Considering this point, we rewrite the exchange-correlation functional as follows:

$$E_{xc}[\rho(r)] = \int dr \rho(r) \epsilon_{xc}([\rho(r)], r). \qquad (3.206)$$

Here, $\epsilon_{xc}([\rho(r)], r)$ is called the local exchange-correlation energy.

Local Density Approximation

In LDA, $\rho(r)$ is approximated as being locally a constant, and hence $\epsilon_{xc}([\rho(r)], r)$ becomes a *function* of the density:

$$E_{xc}[\rho(r)] = \int dr \rho(r) \epsilon_{xc}^{hom}(\rho(r)), \tag{3.207}$$

Here, $\epsilon_{xc}^{hom}(\rho(r))$ is the local exchange-correlation energy of a homogeneous electron gas.

In this case, the exchange-correlation potential is written as,

$$V_{xc}(r) = \epsilon_{xc}^{hom}(\rho(r)) + \rho(r) \frac{d\epsilon_{xc}^{hom}(\rho(r))}{d\rho(r)}. \tag{3.208}$$

LDA is a quite bold approximation. However, the use of LDA is justified a posteriori by its surprising success in predicting physical properties of real materials.

Generalized Gradient Approximation

Although LDA becomes exact in the homogeneous limit, it appears to be a bad approximation in cases with a strongly-varying density. To overcome this problem to some extent, GGA [45] takes account of the spatial gradient of the electron density, which is known to be more consistent with experiments than LDA in many cases.

In GGA, ϵ_{xc} is approximated as a function depending on the density $\rho(r)$ and the gradient of the density $\nabla \rho(r)$. Due to this, GGA has some parameters. Here we introduce probably the most widely used functional among solid state physics, the Perdew–Burke–Ernzerhof parametrization (PBE-GGA) [46].

In PBE-GGA, E_{xc}^{GGA} is divided into two terms, the exchange energy $E_x^{PBE-GGA}$ and the correlation energy $E_c^{PBE-GGA}$:

$$E_x^{PBE-GGA} = \int dr \rho(r) \epsilon_x^{hom}(\rho(r)) F_x(s), \tag{3.209}$$

$$E_c^{PBE-GGA} = \int dr \rho(r) \left[\epsilon_c^{hom}(\rho(r)) + H(t) \right]. \tag{3.210}$$

Here, $\epsilon_{x(c)}^{hom}(\rho(r))$ is the exchange (correlation) energy of a homogeneous electron gas. Let us define dimensionless quantities $s = \nabla \rho / k_F$ and $t = |\nabla \rho| / 2\phi k_{TF} \rho$ along with $\phi = \left[(1+\zeta)^{2/3} + (1-\zeta)^{2/3} \right]/2$, where ζ is the relative spin polarization. Using these parameters,

$$F_x(s) = 1 + \kappa - \frac{\kappa}{1 + \mu s^2 / \kappa}, \tag{3.211}$$

$$H(t) = \frac{e^2}{a_B}\gamma\phi^3\left(1 + \frac{\beta}{\gamma}t^2\frac{1+At^2}{1+At^2+A^2t^4}\right), \quad (3.212)$$

$$A = \frac{\beta}{\gamma}\left[\exp\left(\frac{-\epsilon_c^{\text{hom}}}{\gamma\phi^3\frac{e^2}{a_B}}\right) - 1\right]. \quad (3.213)$$

Parameters $\kappa = 0.804$ and $\mu = 0.21951$ are determined to satisfy Lieb–Oxford bound [47] and to meet the linear response of the spin-unpolarized homogeneous electron gas, respectively. Here k_F is the Fermi wave number, k_{TF} the Thomas-Fermi wave number, e the elementary charge, a_B the Bohr radius, and dimensionless coefficients $\beta = 0.066725$ and $\gamma = (1 - \ln 2)/\pi^2$.

3.5.5 Pseudopotential Method

One can at least numerically obtain electronic properties of actual materials by solving the Kohn–Sham equation with an appropriate density functional. However, practically, a problem lies in the choice of the basis function in solving the Kohn–Sham equation by a computer. Since computational resources are limited, a compact representation for expanding the Kohn–Sham orbitals without sacrificing the numerical accuracy is desired.

Here let us consider expanding the eigenfunction ψ in a plane-wave basis set:

$$\psi_k(r) = \sum_K c_K^k \exp\left(i\,(k+K)\cdot r\right), \quad (3.214)$$

where k is a vector in the first Brillouin zone and K is a reciprocal lattice vector. However it is computationally too demanding to solve the Kohn–Sham equation by a computer due to the following reasons. Electrons in solids are divided broadly into two categories: the core electrons which are strongly localized around the nuclei and the itinerant valence electrons. Due to their localized character, the former is not suitable for expanding by plane waves. The latter also is not suitable for the plane wave expansion since their wave functions should be orthogonal to those of the core electrons. Therefore quite a lot of plane waves is necessary to express the Kohn–Sham orbitals. Both of these problems originate from the difficulty in describing the electronic structure around the nuclei.

On the other hand, there would be a view point that since electrons near the nuclei do not play a so important role in chemical bonds or electronic properties of solids, it is sufficient to treat the contribution of them effectively. Based on this, the pseudopotential method is an formalism to replace the core electrons of and its nucleus with an effective potential and make the representation compact. Here we briefly introduce a pseudo-potential which can construct an effective potential non-empirically: the norm-conserving pseudopotentials.

3.5 Density Functional Theory

Norm-Conserving Pseudopotential

The norm-conserving pseudopotentioal is determined so as to satisfy the following conditions proposed by Hamann, Schlüter, and Chiang:

1. Pseudo and real eigenvalues for valence electrons coincide with each other for a selected prototypical atomic configuration.
2. Pseudo and real wave functions coincide with each other outside of a selected core radius R_c (cutoff radius).
3. The logarithmic derivatives of pseudo and real wave functions and their first energy derivatives coincide with each other at R_c.
4. The integrals from 0 to r of pseudo and real wave functions coincide with each other for $R_c < r$ (norm conservation).

Thanks to the fourth condition, the norm conserving pseudopotentials reproduce scattering properties. In the present study, we perform first-principles calculation using the VASP package [48, 49], based on the pseudopotential method and the projector augmented wave method [50] which is a kind of the generalization of the pseudopotential method. Specifically, we use this to theoretically determine the crystal structure by a structural optimization calculation or to adopt the virtual crystal approximation to take account of the effect of a elemental substitution. The virtual crystal approximation is a method to treat a partial elemental substitution by interpolating potentials of atoms which partially substituted, which hold the original unit cell and hence is computationally cheap.

3.5.6 Augmented Plane Wave Method

The pseudopotential method attempts to overcome problems due to the difficulty in describing the electronic structure near nuclei by introducing an effective potential. As another approach, here we introduce the augmented plane wave (APW) method and some generalizations of it, which describes wave functions by atomic-like functions close to the nuclei and plane waves in the region far away from the nuclei [51]. As its generalizations, there are the linearized APW (LAPW) [52], LAPW+LO (local orbital) [53], and APW+lo [54] method.

APW Basis Function

In the APW method, assuming a sphere of radius R_α (muffin-tin sphere) to divide space into two regions: the core region S_α and outside I. The wave functions are expanded as:

$$\phi_K^k(r, E) = \begin{cases} \frac{1}{\sqrt{V}} e^{i(k+K)\cdot r}, & (r \in I), \\ \sum_{l,m} A_{lm}^{\alpha,k+K} u_l^\alpha(r', E) Y_{lm}(r'), & (r \in S_\alpha). \end{cases} \quad (3.215)$$

Here, $u_l^\alpha(r', E) Y_{lm}(r')$ is an atomic wave function and r' is the distance between the nucleus and r. The coefficient $A_{lm}^{\alpha,k+K}$ is determined so that the wave function is continuous at the sphere boundary.

However, this expression is practically somewhat problematic since the basis functions depend on the eigenvalues. This makes the problem unable to solve as a simple eigenvalue problem. Therefore it is necessary to contrive to resolve this issue, such as linearizing around a proper energy E_0.

LAPW(+LO) Method

In the LAPW method, we employ the Taylor expansion of the APW basis functions around a certain energy E_0:

$$\phi_K^k(r) = \begin{cases} \frac{1}{\sqrt{V}} e^{i(k+K)\cdot r}, & (r \in I), \\ \sum_{l,m} \left[A_{lm}^{\alpha,k+K} u_l^\alpha(r', E_0^\alpha) + B_{lm}^{\alpha,k+K} \dot{u}_l^\alpha(r', E_0^\alpha) \right] Y_{lm}(r'), & (r \in S_\alpha), \end{cases}$$
$$(3.216)$$

where $\dot{u}$ is the energy derivative of u. There is not a universal guiding principle to choose E_0, but usually we choose angular-momentum-dependent $E_{1,l}^\alpha$. Since the energy difference is unknown, an undetermined parameter $B_{lm}^{\alpha,k+K}$ has to be introduced. $A_{lm}^{\alpha,k+K}$ and $B_{lm}^{\alpha,k+K}$ are determined so that the wave function and its derivative is continuous at the sphere boundary.

Since $E_{1,l}^\alpha$ depends only on the angular momentum l, this can not be determined uniquely, if states with the same l but different principal quantum number are both valence states (e.g. $4p$ and $3p$ states of bcc Fe). To treat this kind of cases, in the LAPW+LO method, another type of basis function is added to consider the effect of the local orbitals in the muffin-tin sphere:

$$\phi_{\alpha,\text{LO}}^{lm}(r) = \begin{cases} 0, & (r \in I), \\ \left[A_{lm}^{\alpha,\text{LO}} u_l^\alpha(r', E_{1,l}^\alpha) + B_{lm}^{\alpha,\text{LO}} \dot{u}_l^\alpha(r', E_{1,l}^\alpha) + C_{lm}^{\alpha,\text{LO}} u_l^\alpha(r', E_{2,l}^\alpha) \right] Y_{lm}(r'), & (r \in S_\alpha). \end{cases}$$
$$(3.217)$$

Here, $E_{1,l}$ and $E_{2,l}$ are energy of the orbitals contributing to the valence state. The coefficients $A_{lm}^{\alpha,\text{LO}}$, $B_{lm}^{\alpha,\text{LO}}$, and $C_{lm}^{\alpha,\text{LO}}$ is determined by requiring the normalization condition and zero value of LO and its slope.

3.5 Density Functional Theory

APW(+lo) Method

It is possible to remove the energy dependence of the basis function also from the APW basis function by introducing the local orbitals, which is called the APW+lo method.

The APW+lo basis function is described by the adding the local orbital function only in the muffin-tin sphere,

$$\phi_{\alpha,\text{lo}}^{lm}(r) = \begin{cases} 0, & (r \in I), \\ \left[A_{lm}^{\alpha,\text{lo}} u_l^\alpha(r', E_{1,l}^\alpha) + B_{lm}^{\alpha,\text{lo}} \dot{u}_l^\alpha(r', E_{1,l}^\alpha) \right] Y_{lm}(r'), & (r \in S_\alpha). \end{cases} \quad (3.218)$$

to the APW wave function with a fixed energy,

$$\phi_K^k(r, E) = \begin{cases} \frac{1}{\sqrt{V}} e^{(k+K) \cdot r}, & (r \in I), \\ \sum_{l,m} A_{lm}^{\alpha, k+K} u_l^\alpha(r', E_{1,l}^\alpha) Y_{lm}(r'), & (r \in S_\alpha), \end{cases} \quad (3.219)$$

The coefficients $A_{lm}^{\alpha,\text{lo}}$ and $B_{lm}^{\alpha,\text{lo}}$ are determined by requiring the normalization condition and zero value if LO is on the muffin-tin sphere.

Choosing an appropriate energy, the computational time of the APW+lo calculations is shorter than that of the LAPW method. On the other hand, it requires considerably larger basis than the LAPW basis set for the same level of the accuracy. Therefore it is better to add extra basis functions only if necessary. This leads to the calculation with a mixed LAPW/APW+lo basis set. In the present study, we perform first-principles calculations with this method using the WIEN2K package [55]. Specifically, we use this method to calculate the band structures and the density of states. In this method, one important parameter defining the accuracy of a calculation is RK_{max}: the product the smallest muffin-tin sphere radius R and the largest K-vector K_{max} of the plane wave expansion of the wave function, which determines the size of the basis set.

3.6 Model Construction from DFT Results

It is difficult to treat strong electron correlation within DFT. For instance, it is known that the widely used LDA/GGA functional fails to describe physics originating from strong correlation effects, exampled by the Mott transition. Even if one can treat strong electron correlation in an efficient way, it is still difficult to find out an essential part of electronic properties from enormous amount of information. Regarding these points, constructing an effective model and analyzing it based on the many-body theory are useful in treating electron correlation or find out a physical picture. In this section, we describe a method to construct an effective model from the DFT calculation exploiting maximally-localized Wannier functions proposed by Marzari and Vanderbilt [56, 57].

3.6.1 Maximally-Localized Wannier Functions

As we mentioned above, the Wannier functions have arbitrariness in the choice of the unitary matrix U^k. Taking an advantage of this property, the maximally-localized Wannier function method enables us to refine iteratively this arbitrary degrees of freedom so that they are well localized around their center. This is a powerful tool to construct a minimal model which accurately reproduces the low-energy electronic structure in terms of atomic-like orbitals, without a strong bias.

Here we introduce the second moment around the center of the Wannier functions Ω as a measure of their spatial spread:

$$\Omega = \sum_n \left[\langle r^2 \rangle_n - \langle r \rangle_n^2 \right] \tag{3.220}$$

$$= \sum_n \left[\langle w_{0n} | r^2 | w_{0n} \rangle - |\langle w_{0n} | r | w_{0n} \rangle|^2 \right]. \tag{3.221}$$

Decomposing Ω into the unitary-invariant part Ω_I and the non-unitary-invariant part $\tilde{\Omega}$,

$$\Omega = \Omega_\mathrm{I} + \tilde{\Omega} = \Omega_\mathrm{I} + \Omega_\mathrm{OD} + \Omega_\mathrm{D}, \tag{3.222}$$

where,

$$\Omega_\mathrm{I} = \sum_n \left[\langle w_{0n} | r^2 | w_{0n} \rangle - \sum_{R_m} |\langle w_{R_m n} | r | w_{0n} \rangle|^2 \right]$$

$$= \sum_{\alpha=x,y,z} \mathrm{Tr} \{ P r_\alpha Q r_\alpha \}, \tag{3.223}$$

$$\tilde{\Omega} = \sum_n \sum_{Rm \neq 0n} |\langle w_{Rm} | r | w_{0n} \rangle|^2, \tag{3.224}$$

$$\Omega_\mathrm{OD} = \sum_{m \neq n} \sum_R |\langle w_{Rm} | r | w_{0n} \rangle|^2, \tag{3.225}$$

$$\Omega_\mathrm{D} = \sum_n \sum_{R \neq 0} |\langle w_{Rn} | r | w_{0n} \rangle|^2. \tag{3.226}$$

Here P is a projection operator $P = \sum_{Rn} |w_{Rn}\rangle \langle w_{Rn}|$ and $Q = 1 - P$. Since Ω_I is invariant under any arbitrary unitary transformation of the Bloch orbitals, it is sufficient to minimize $\tilde{\Omega}$ in order to find the minimum value of Ω. To ensure the physical

3.6 Model Construction from DFT Results

meaning of the result, an initial guess for the maximally localized Wannier functions is projected using the atomic orbitals. One can obtain the tight-binding model by transforming the Bloch functions to the Wannier functions using the resulting $U^{(k)}$.

In the present study, we construct the effective tight-binding model by this method using the WANNIER90 [58] and WIEN2WANNIER [59] codes.

References

1. Abrikosov AA, Gorkov LP, Dzyaloshinski IE (1975) Methods of quantum field theory in statistical physics. Courier Corporation
2. Fetter AL, Walecka JD (2003) Quantum theory of many-particle systems. Courier Corporation
3. Takada Y (1999) Tatai mondai (written in Japanese). Asakura Publishing
4. Kuroki K, Aoki H (1999) Tatai Densironn II Choudendou (written in Japanese). University of Tokyo Press
5. Luttinger JM, Ward JC (1960) Phys Rev 118:1417
6. Baym G, Kadanoff LP (1961) Phys Rev 124:287
7. Bardeen J, Cooper LN, Schrieffer JR (1957) Phys Rev 108:1175
8. Bickers NE, Scalapino DJ, White SR (1989) Phys Rev Lett 62:961
9. Bickers NE, White SR (1991) Phys Rev B 43:8044
10. Metzner W, Vollhardt D (1989) Phys Rev Lett 62:324
11. Georges A, Kotliar G (1992) Phys Rev B 45:6479
12. Georges A, Kotliar G, Krauth W, Rozenberg MJ (1996) Rev Mod Phys 68:13
13. Hettler MH, Tahvildar-Zadeh AN, Jarrell M, Pruschke T, Krishnamurthy HR (1998) Phys Rev B 58:R7475
14. Lichtenstein AI, Katsnelson MI (2000) Phys Rev B 62:R9283
15. Kotliar G, Savrasov SY, Pálsson G, Biroli G (2001) Phys Rev Lett 87:186401
16. Toschi A, Katanin AA, Held K (2007) Phys Rev B 75:045118
17. Rubtsov AN, Katsnelson MI, Lichtenstein AI (2008) Phys Rev B 77:033101
18. Brener S, Hafermann H, Rubtsov AN, Katsnelson MI, Lichtenstein AI (2008) Phys Rev B 77:195105
19. Otsuki J, Hafermann H, Lichtenstein AI (2014) Phys Rev B 90:235132
20. Taranto C et al (2014) Phys Rev Lett 112:196402
21. Gukelberger J, Huang L, Werner P (2015) Phys Rev B 91:235114
22. Kitatani M, Tsuji N, Aoki H (2015) Phys Rev B 92:085104
23. Kitatani M, Tsuji N, Aoki H (2017) Phys Rev B 95:075109
24. Altland A, Simons BD (2010) Condensed matter field theory. Cambridge University Press
25. Janiš V, Vollhardt D (1992) Int J Mod Phys B 6:731
26. Kajueter H, Kotliar G (1996) Phys Rev Lett 77:131
27. Potthoff M, Wegner T, Nolting W (1997) Phys Rev B 55:16132
28. Arsenault L-F, Sémon P, Tremblay A-MS (2012) Phys Rev B 86:085133
29. Rubtsov AN, Savkin VV, Lichtenstein AI (2005) Phys Rev B 72:035122
30. Werner P, Comanac A, de' Medici L, Troyer M, Millis AJ (2006) Phys Rev Lett 97:076405
31. Gull E et al (2011) Rev Mod Phys 83:349
32. Shinaoka H, Gull E, Werner P (2017) Comput Phys Commun 215:128
33. Boehnke L, Hafermann H, Ferrero M, Lechermann F, Parcollet O (2011) Phys Rev B 84:075145
34. Gaenko A et al (2017) Comput Phys Commun 213:235
35. Bauer B et al (2011) J Stat Mech: Theory Exp 2011:P05001
36. Yoshida T et al (2003) Phys Rev Lett 91:027001
37. Yoshida T et al (2009) Phys Rev Lett 103:037004
38. Hafermann H et al (2009) Phys Rev Lett 102:206401

39. Cottenier S (2002–2013) DFT and the family of (L)APW-methods: a step-by-step introduction
40. Takada Y (2009) Tatai Mondai Tokuron (written in Japanese). Asakura Publishing
41. Martin RM (2004) Electronic structure: basic theory and practical methods. Cambridge University Press, Cambridge
42. Hohenberg P, Kohn W (1964) Phys Rev 136:B864
43. Kohn W, Sham LJ (1965) Phys Rev 140:A1133
44. Janak JF (1978) Phys Rev B 18:7165
45. Perdew JP, Burke K, Wang Y (1996) Phys Rev B 54:16533
46. Perdew JP, Burke K, Ernzerhof M (1996) Phys Rev Lett 77:3865
47. Lieb EH, Oxford S (1981) Int J Quantum Chem 19:427
48. Kresse G, Furthmüller J (1996) Phys Rev B 54:11169
49. Kresse G, Joubert D (1999) Phys Rev B 59:1758
50. Blöchl PE (1994) Phys Rev B 50:17953
51. Slater JC (1937) Phys Rev 51:846
52. Andersen OK (1975) Phys Rev B 12:3060
53. Takeda T, Kubler J (1979) J Phys F: Metal Phys 9:661
54. Sjöstedt E, Nordström L, Singh D (2000) Solid State Commun 114:15
55. Blaha P, Schwarz K, Madsen G, Kvasnicka D, Luitz J (2001) An augmented plane wave+ local orbitals program for calculating crystal properties
56. Marzari N, Vanderbilt D (1997) Phys Rev B 56:12847
57. Souza I, Marzari N, Vanderbilt D (2001) Phys Rev B 65:035109
58. Mostofi AA et al (2008) Comput Phys Commun 178:685
59. Kuneš J et al (2010) Comput Phys Commun 181:1888

Chapter 4
FLEX+DMFT Analysis for Superconductivity in Systems with Coexisting Wide and Narrow Bands

Abstract In this chapter, we study superconductivity and electron correlation effects in the bilayer Hubbard model with coexisting wide and incipient narrow bands originating from multiple sites within the unit cell. We have applied the FLEX+DMFT method, which we have extended to multi-band systems in the present thesis, to study electron correlation effects and spin-fluctuation-mediated superconductivity. We found that the method can capture the so-called pseudogap behavior, a hallmark of strong electron correlation, which is hardly seen in the FLEX result. Based on the analysis of the Eliashberg equation, we found that a pairing mechanism exploiting an incipient band discussed in the previous studies is still valid even with the local vertex corrections coming from the DMFT part of the self-energy.

Keywords FLEX+DMFT method · Multi-band system · Incipient band

4.1 Motivation

As mentioned in Chap. 2, a possible enhancement of superconductivity due to the interband scattering processes involving an incipient band, namely, a band lying below, but not far away from, the Fermi level has been discussed in the previous studies [1–10]. References [10, 11] suggest that this pairing mechanism can be applied to various systems with coexisting wide and narrow bands in quasi-one-, two-, and even three-dimensions. However, since these proposals are based on the FLEX approximation, it is still unclear how the strong correlation effects affect this scenario. A method incorporating both local and non-local (e.g. magnetic/charge fluctuations) correlations is necessary to treat this kind of pairing mechanisms since we have to consider a pairing interaction mediated by non-local (magnetic) correlations in the presence of the strong local correlation.

A study on the Hubbard model on a diamond chain with ED and DMRG [12] suggests an enhancement of superconductivity in a system with wide and flat bands. Although these methods are numerically exact, the study has been performed only for a quite specific situation, i.e. a purely one-dimensional lattice, at zero temperature, and with one of the bands being perfectly flat. A DCA study [13] has shown that superconductivity is strongly enhanced in the bilayer Hubbard model with two identical, bonding and antibonding bands, which is also a specific situation. Therefore a systematic analysis on the role of strong correlations due to the strong interband scattering is highly desired.

Given this background, we study the bilayer Hubbard model for various cases. To treat both local and non-local electron correlations self-consistently with a moderate computational cost, we adopt the multi-band extension of the FLEX+DMFT method, which we have formulated in Chap. 3. Specifically, we address the problem of whether the pairing mechanism exploiting an incipient band is still valid or not in the presence of strong local correlation in various systems.

4.2 Formulation

4.2.1 Model

In the present study, we consider the bilayer Hubbard model shown in Fig. 4.1. Its Hamiltonian is given by Eq. (3.165), where indices α and β ($=1, 2$) indicate a layer. The hopping term can be rewritten in the momentum space as

$$\begin{pmatrix} c^{1\dagger}_{k\sigma} & c^{2\dagger}_{k\sigma} \end{pmatrix} \hat{\mathcal{H}}(k) \begin{pmatrix} c^{1}_{k\sigma} \\ c^{2}_{k\sigma} \end{pmatrix} \tag{4.1}$$

where

$$\hat{\mathcal{H}}(k) = \begin{pmatrix} 2(t_x \cos k_x + t_y \cos k_x) & 2(t'_x \cos k_x + t'_y \cos k_x) + t_d \\ 2(t'_x \cos k_x + t'_y \cos k_x) + t_d & 2(t_x \cos k_x + t_y \cos k_x) \end{pmatrix}. \tag{4.2}$$

One can obtain the bare band dispersion $E_\pm(k)$ by diagonalizing $\hat{\mathcal{H}}(k)$

$$E_\pm = 2\left[(t_x \pm t'_x)\cos k_x + (t_y \pm t'_y)\cos k_y\right] \pm t_d \tag{4.3}$$

with the eigenvectors u_1 and u_2

$$u_1 = \frac{1}{\sqrt{2}}\begin{pmatrix} 1 \\ 1 \end{pmatrix},$$

$$u_2 = \frac{1}{\sqrt{2}}\begin{pmatrix} 1 \\ -1 \end{pmatrix}. \tag{4.4}$$

It can be seen that $t'_{x(y)}$ controls the band width along the $k_{x(y)}$ direction and the ratio $t_y^{(\prime)}/t_x^{(\prime)}$ the dimensionality. Namely, the system has a perfectly flat band if $t_x = t_y = t'_x = t'_y = t$, and the system becomes purely one-dimensional if $t_y = t'_y = 0$. Hereafter we use t as the unit of energy. In the present study, we have employed several

4.2 Formulation

(a)

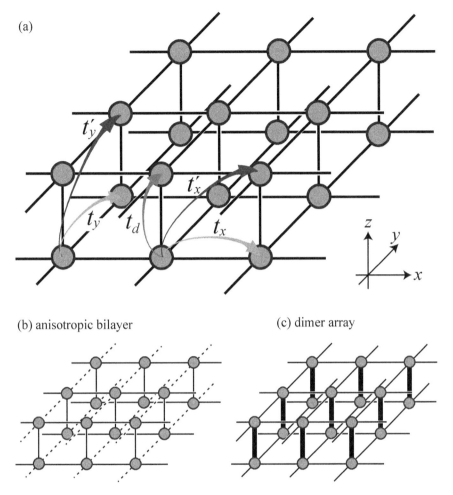

(b) anisotropic bilayer   (c) dimer array

Fig. 4.1 a A schematic of the bilayer model. Arrows indicate hopping integrals. **b** A schematic of the anisotropic bilayer model, where the dashed lines represent a weak coupling toward the y direction. **c** A schematic of the dimer array model, where the thick lines represent a strong intra-dimer coupling

sets of hopping parameters listed in Table 4.1. (a) corresponds to the case of coexisting wide and flat bands with square lattice symmetry, (b) to the case of coexisting wide and narrow bands with square lattice symmetry, (c) to the case of coexisting wide and narrow bands in quasi-one-dimension and roughly corresponding to the hopping parameters of the model for $Sr_3Mo_2O_7$ with the experimental structure, which we will discuss in Chap. 5, and (d) to the case of two identical bands with bonding-antibonding splitting $2t_d$ with square lattice symmetry. Hereafter we refer to these models as (a) the wide-and-flat-band model, (b) the wide-and-narrow-band model, (c) the anisotropic bilayer model, and (d) the dimer-array model (as in Ref. [14],

Table 4.1 Hopping parameters used in the present study. The fifth column of (d) means t_d is treated as a parameter in the present study

	t_x/t	t_y/t	t'_x/t	t'_y/t	t_d/t
(a)	1	1	1	1	1
(b)	1	1	0.5	0.5	1
(c)	−1	−0.15	0.25	0	0.85
(d)	1	1	0	0	t_d/t

and as schematically shown in Fig. 4.1c), or more simply (a–d), respectively. Note that the anisotropic bilayer model can be considered as consisting of two-leg ladders extended in the x direction coupled weakly in the y direction as schematically shown in Fig. 4.1b. The band structures and the density of states for these choices are shown in Figs. 4.2 and 4.3, respectively. In the many-body analysis, we vary the band filling n for (a–c) and the interlayer coupling strength t_d for (d). Since the dimer array model consists of two bands with identical band width, one may consider that this model cannot be categorized as a system with coexisting wide and narrow bands. Still, we will give a reason why this may also be considered as a system with coexisting wide and narrow bands, once the electron correlation is considered.

4.2.2 Many-Body Analysis

In the present study, we adopt the FLEX+DMFT method to treat the electron correlation effect. In order to treat interlayer and intralayer non-local correlations on an equal footing, we neglect the intersite corrections in the DMFT part of the self-energy. Therefore we restrict ourselves to intermediate interaction strength to justify this simplification. As seen in the Chap. 5 and confirmed later, the diagonal part of the gap function reverses its sign around the optimal filling in many cases, so we also omit the DMFT kernel for the Eliashberg equation. We set the on-site interactions as $U/t = 5$ and the temperature is fixed at $\beta t = 14$ ($k_B T/t \sim 0.07$), and we take 32 × 32 k-meshes and 1024 Matsubara frequencies in total. For the dimer array model, the band filling is fixed at $n = 2.40$. For the DMFT solver, we employ the CT-HYB using codes based on the ALPSCore libraries and ALPSCore/CT-HYB [15–17] since perturbative solvers do not work so well unlike in single-band cases. We set the number of Legendre polynomials for the expansion of the Green's function n_{Leg} as 50. We have confirmed the results are almost unchanged even if we employ 64 × 64 k-meshes, 2048 Matsubara frequencies and $n_{Leg} = 70$ as described in Appendix B. To obtain the spectral function, we perform the analytic continuation with the maximum entropy method using the ΩMaxent code [18]. In estimating statistical errors coming from the CT-HYB solver, we have performed another 20 iterations after the self-consistent loop is converged and the standard deviation of them is used as

4.2 Formulation

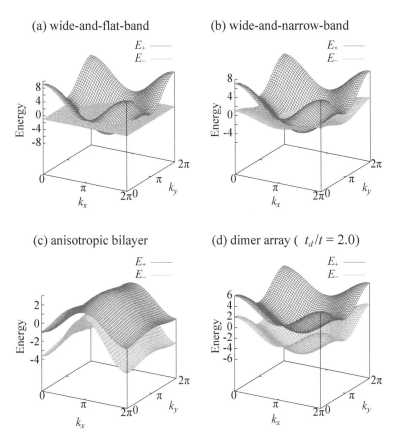

Fig. 4.2 Band structures for **a** the wide-and-flat-band model, **b** the wide-and-narrow-band model, **c** the anisotropic bilayer model, and **d** the dimer-array model in which we take $t_d/t = 2$

a statistical error. For some parameters, only 15–19 iterations are performed after convergence since the negative sign problem becomes serious due to the numerical errors.

To be strict, since we treat two-dimensional systems, the Mermin–Wagner theorem [19] prohibits the emergence of the superconducting phase transition at finite temperature. Finite T_c (namely $\lambda > 1$) is obtained in the present formalism because a mean-field approximation is used in the derivation of the Eliashberg equation. However, it has been shown that a weak but finite three-dimensionality does not appreciably change T_c's within the FLEX approximation [20]. Since we employ the same formalism for the superconductivity, the situation can be considered as similar to the FLEX case. Hence, T_c's (or λ's) in the present study should be considered as those for the three-dimensional systems in which the present lattices are weakly coupled in the z direction.

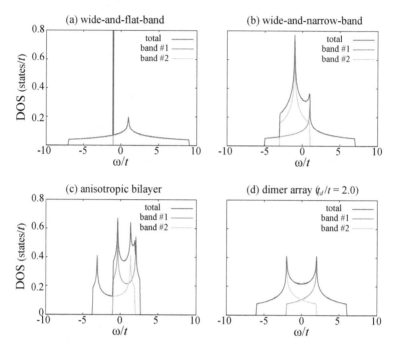

Fig. 4.3 Non-interacting density of states for **a** the wide-and-flat-band model, **b** the wide-and-narrow-band model, **c** the anisotropic bilayer model, and **d** the dimer-array model in which we take $t_d/t = 2$

4.3 Results and Discussion

4.3.1 Superconductivity

We first display the band filling/interlayer coupling strength dependence of the eigenvalue λ of the linearized Eliashberg equation and the Stoner factor α_S for each model in Figs. 4.4, 4.5, 4.6 and 4.7. For comparison, we also display the FLEX results for the same parameters. First of all, it can be seen in all the cases that superconductivity is optimized for parameter values where the Stoner factor is somewhat reduced from its maximum value, which is similar to the FLEX result. The optimal parameters are $n = 2.26$ for the wide-and-flat-band model, $n = 2.36$ for the wide-and-narrow-band model, $n = 1.32$ for the anisotropic bilayer model, and $t_d/t = 1.75$ for the dimer array model. This qualitative consistency between FLEX+DMFT and FLEX indeed shows that the wide applicability of the pairing mechanism exploiting coexisting wide and narrow bands discussed within the FLEX approximation [10, 11] is still, at least qualitatively, valid even when the strong local correlation effects are taken into account. Note that the enhancement of α_S around $n = 3$ in the wide-and-narrow-band model ($n = 1.0$ in the anisotropic bilayer model) is due to the intraband scatterings,

4.3 Results and Discussion

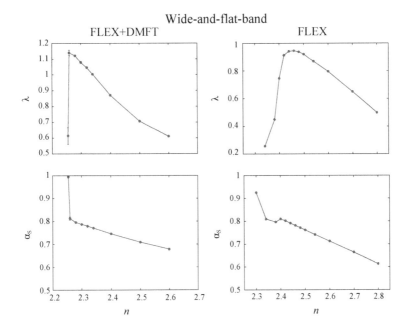

Fig. 4.4 (Left panel) FLEX+DMFT results of the band filling dependence of (upper panel) the eigenvalue λ of the linearized Eliashberg equation and (lower panel) the Stoner factor for the wide-and-flat-band model. For comparison, the FLEX results are also displayed in the right panels

which originates from the one band being nearly half-filled and the other being nearly fully-filled (empty), so that the system can be regarded as an effective single-band system.

If we compare the FLEX+DMFT and the FLEX results more quantitatively, there are indeed some differences. We notice that the optimized value of λ is larger in FLEX+DMFT than in FLEX. This is a tendency which was already seen in the single band Hubbard model [21]. If we compare the value of α_S for the parameter sets where λ is optimized for each cases, we find that FLEX+DMFT gives larger values than FLEX. In understanding the reasons for that, it is important to note that whereas the FLEX approximation overestimates the local part of the self-energy, the FLEX+DMFT method overestimates the non-local part of the self-energy as pointed out in the previous studies [21, 22]. If we compare λ for both methods at similar α_S, namely for similar strengths of the non-local self-energy, we can see that the FLEX approximation gives strongly reduced λ compared to the FLEX+DMFT. In addition, except for the above-mentioned effective single-band cases, α_S is not so different between FLEX and FLEX+DMFT when compared at the same parameter values. Therefore the parameter values for which superconductivity is optimized in these cases is mainly determined by the local self-energy, which is overestimated in the FLEX approximation.

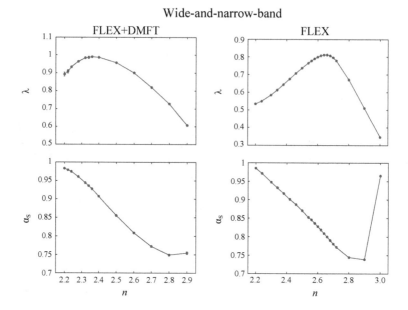

Fig. 4.5 (Left panel) FLEX+DMFT results of the band filling dependence of (upper panel) the eigenvalue λ of the linearized Eliashberg equation and (lower panel) the Stoner factor for the wide-and-narrow-band model. For comparison, the FLEX results are also displayed in the right panels

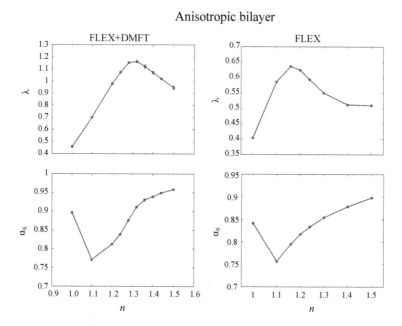

Fig. 4.6 (Left panel) FLEX+DMFT results of the band filling dependence of (upper panel) the eigenvalue λ of the linearized Eliashberg equation and (lower panel) the Stoner factor for the anisotropic bilayer model. For comparison, the FLEX results are also displayed in the right panels

4.3 Results and Discussion

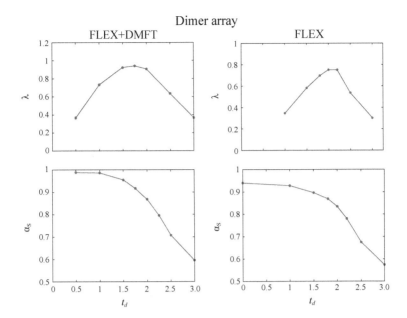

Fig. 4.7 (Left panel) FLEX+DMFT results of the interlayer coupling strength dependence of (upper panel) the eigenvalue λ of the linearized Eliashberg equation and (lower panel) the Stoner factor for the dimer array model. For comparison, the FLEX results are also displayed in the right panels

The dimer array model has been studied within DCA [13]. Since we employ different temperature, interaction strength, and band filling to those of the DCA study in Ref. [13], we can not make a direct comparison. Still, in the DCA study, the superconducting susceptibility is strongly enhanced for $t_d/t \sim 2$, where the interlayer pairing interaction and the interlayer nearest neighbor spin fluctuation attains maximum values. This is also consistent with our result shown in Fig. 4.7.

In Figs. 4.8, 4.9, 4.10 and 4.11, we depict the anomalous self-energy at the lowest Matsubara frequency for parameter sets for which the superconductivity is optimized in each model. Here Δ_{11} is the intralayer component, Δ_{12} the interlayer component. In obtaining the band representation of the anomalous self-energy, we perform the unitary transformation which diagonalizes $\hat{\mathcal{H}}(k)$. $\Delta_{\text{band}1(2)}$ corresponds to the component represented by the eigenvector $u_{1(2)}$, namely

$$\Delta_{\text{band}1} = \Delta_{11} + \Delta_{12}$$
$$\Delta_{\text{band}2} = \Delta_{11} - \Delta_{12}. \qquad (4.5)$$

It can be clearly seen that the obtained anomalous self-energies indicate the s-wave gap symmetry having opposite sign between the two bands but without sign changing within each band, and the interlayer component Δ_{12} has the dominant contribution,

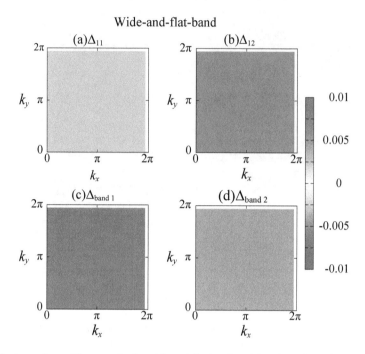

Fig. 4.8 Anomalous self-energy for the wide-and-flat-band model at the optimal filling $n = 2.26$. Panels **a** and **b** show the intralayer component Δ_{11} and **b** the interlayer component Δ_{12}. In lower panels, the anomalous self-energy in the band representation **c** for the band 1 $\Delta_{11} + \Delta_{12}$ and **d** for the band 2 $\Delta_{11} - \Delta_{12}$

which is consistent with those obtained with the FLEX approximation and the DCA calculation.

As mentioned above, we omit the DMFT kernel for the Eliashberg equation in the present study, but it can give rise to some corrections to certain extent unlike in the d-wave pairing case, where the DMFT contribution completely vanishes by the momentum summation. Let us examine the DMFT correction for the anomalous self-energy. Note that since we omit the intersite DMFT corrections for the normal self-energy, here we consider only the on-site correction that for the anomalous part. For the wide-and-narrow-band model, the anisotropic bilayer model, and the dimer array model, Δ_{11} changes its sign in the Brillouin zone like $s_{\pm}$-wave symmetry as seen in Figs. 4.9a, 4.10 and 4.11a. Since the DMFT correction has to be *momentum independent*, a cancellation should occur in the summation over momenta in the linearized Eliashberg equation. Therefore we consider the effect of the DMFT kernel for the linearized Eliashberg equation to be small in these cases. On the other hand, Δ_{11} does not change its sign in the wide-and-flat-band model (Fig. 4.4), so that a cancellation does not occur. However, since the interlayer pairing plays a dominant role in the present case, we expect the DMFT correction would give rise to only quantitative changes.

4.3 Results and Discussion

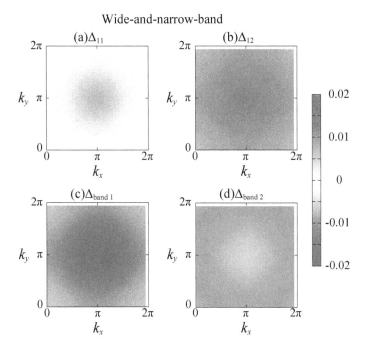

Fig. 4.9 Anomalous self-energy for the wide-and-narrow-band model at the optimal filling $n = 2.36$. Panels **a** and **b** show the intralayer component Δ_{11} and **b** the interlayer component Δ_{12}. In lower panels, the anomalous self-energy in the band representation **c** for the band 1 $\Delta_{11} + \Delta_{12}$ and **d** for the band 2 $\Delta_{11} - \Delta_{12}$

4.3.2 Spectral Function

As seen above, the FLEX+DMFT method gives qualitatively consistent results with those obtained within the FLEX approximation and the dynamical cluster approximation with respect to superconductivity, namely the DMFT part of the self-energy gives rise to quantitative corrections. Then, how are the spectral functions, namely, the renormalized density of states? Hereafter, we concentrate on the wide-and-narrow-band model and the anisotropic bilayer model. The reason for this is as follows: (1) the wide-and-flat-band model has the strongly singular density of states originating from the flat band, which makes the analytic continuation quite difficult, and (2) two models we are focusing on here are typical among the models treated in the present study.

The resulting spectral functions for the wide-and-narrow-band model and for the anisotropic bilayer model are shown in Figs. 4.12 and 4.13, respectively. For comparison, the spectral functions obtained with the FLEX approximation, which also have not been discussed so far in the previous studies, are also shown. In each figure, we show the spectral functions for three typical band fillings: (1) a too small band filling, (2) the optimal band filling, and (3) a too large band filling. In other

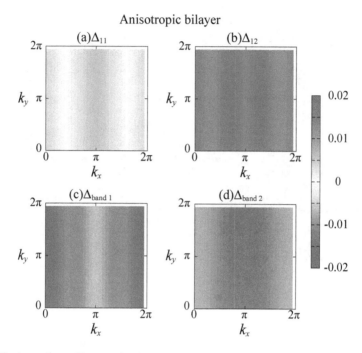

Fig. 4.10 Anomalous self-energy for the anisotropic bilayer model at the optimal filling $n = 1.32$. Panels **a** and **b** show the intralayer component Δ_{11} and **b** the interlayer component Δ_{12}. In lower panels, the anomalous self-energy in the band representation **c** for the band 1 $\Delta_{11} + \Delta_{12}$ and **d** for the band 2 $\Delta_{11} - \Delta_{12}$

words, for the wide-and-narrow-band model, a too small band filling n corresponds to a too large Stoner factor α_S and a too large n corresponds to a too small α_S, and for the anisotropic bilayer model, vice versa. For both models, there are common features in the spectral functions obtained with FLEX+DMFT and that obtained with FLEX. For a too small Stoner factor, the narrow band does not have weight at the zero energy (i.e. the chemical potential), whereas the wide band is not strongly renormalized. This implies that the interband pairing interaction is weak and hence superconductivity is degraded. On the other hand, for a too large Stoner factor, we can see that the both bands have considerably large weight at the chemical potential, and the wide band is strongly renormalized. This leads to the suppression of superconductivity. The spectral functions have an intermediate feature: the density of states of the narrow band is not large at the chemical potential, but considerably large at just above the chemical potential while the wide band is not so strongly renormalized. Namely, the narrow band can be regarded as an incipient band.

A difference can be seen between the spectral function obtained with FLEX+DMFT and that obtained with FLEX: the former is strongly band-filling dependent, while the latter not so much. For small α_S, the wide band is not strongly renormalized and the narrow band sits away from the chemical potential. On the other hand, for

4.3 Results and Discussion

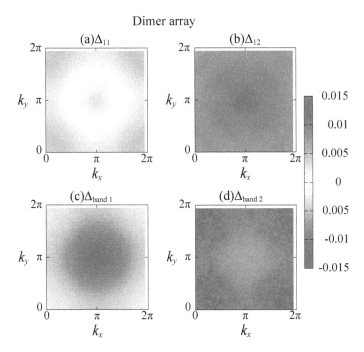

Fig. 4.11 Anomalous self-energy for the dimer array model at the optimal coupling strength $t_d/t = 1.75$. Panels **a** and **b** show the intralayer component Δ_{11} and **b** the interlayer component Δ_{12}. In lower panels, the anomalous self-energy in the band representation **c** for the band 1 $\Delta_{11} + \Delta_{12}$ and **d** for the band 2 $\Delta_{11} - \Delta_{12}$

large α_S both bands are strongly renormalized and a clear pseudogap behavior, a dip in the spectral function, emerges at the chemical potential, which is a hallmark of strong correlations. Since the pseudo gap behavior can be seen in both bands, it would originate from strong interband scatterings. This behavior is hardly seen in the FLEX results. From these results, we can say that the present formalism can capture physics arising from the strong local correlation effect.

Another important observation is the asymmetric renormalization of the narrow band seen in both FLEX+DMFT and FLEX results. Namely, the spectral weight is concentrated in the low energy region by the interaction, in contrast to the non-interacting case, where each band is completely symmetric with respect to the band center. This can be interpreted as follows: since correlation effects seem to more easily take place in the low-energy part rather than in the high-energy part, the spectral weight would tend to be concentrated in the low energy. This gives rise to the large density of states at the narrow band edge even in the two-dimensional system, where the non-interacting density of states is not so large around the band edge. Namely, the density of states of two-dimensional systems would act as that of an effectively quasi-one-dimensional system due to correlation effects, which seems to work in favor of a pairing interaction exploiting an incipient band. We can expect

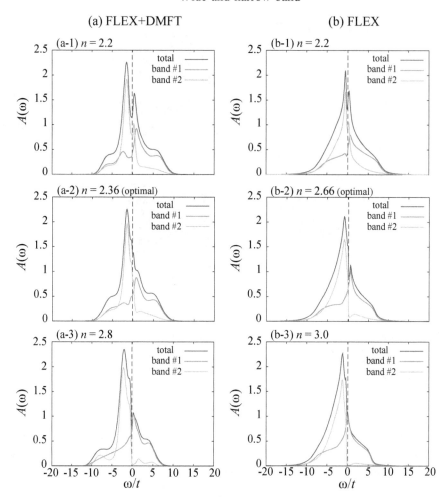

Fig. 4.12 (Left panels) FLEX+DMFT result of the spectral function of the wide-and-narrow-band model for (a–1) a too small band filling, (a–2) the optimal band filling, and (a–3) a too large band filling. For comparison, the FLEX result is also displayed in the right panels (b–1–3). The vertical dashed line represents the chemical potential

this leads to an intuitive understanding for an enhancement of superconductivity in the bilayer Hubbard model over a wide parameter range including the dimer array model, where the width of the two bands are completely the same.

Since the maximum entropy method gives an inference for the spectral function in a statistical manner, we have to be careful of whether the obtained spectra are physically meaningful ones. To address this problem, we also perform the analytic continuation using the Padé approximation for the band represented spectral function

4.3 Results and Discussion

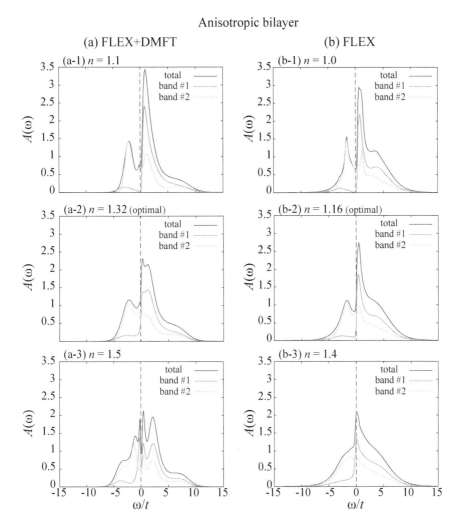

Fig. 4.13 (Left panels) FLEX+DMFT result of the spectral function of the anisotropic bilayer model for (a–1) a too small band filling, (a–2) the optimal band filling, and (a–3) a too large band filling. For comparison, the FLEX result is also displayed in the right panels (b–1–3). The vertical dashed line represents the chemical potential

of the anisotropic bilayer model with $n = 1.32$, in which a quite strong pseudogap is observed. As discussed in Appendix C, we have confirmed that the obtained spectral function does not have an appreciable change, which strongly suggest the present result is not an artifact of the analytic continuation method.

4.4 Summary

To summarize, we have studied unconventional superconductivity and electron correlation effects in the bilayer Hubbard model exploiting a multi-band extension of the FLEX+DMFT method. Based on the analysis of the linearized Eliashberg equation, the pairing mechanism exploiting coexisting wide and narrow bands, which has been mainly discussed using the FLEX approximation in the previous studies is still valid in the presence of vertex corrections coming from the DMFT part of the self-energy. Also we have confirmed that when the width of the non-interacting bands is completely the same, the present formalism gives qualitatively consistent results with DCA. In addition, interestingly, it is found that the present formalism can capture the pseudogap behavior in the spectral function, a fingerprint of the strong interband correlation. We also pointed out that there can be seen the asymmetric renormalization of the narrow band in both FLEX+DMFT and FLEX results, which may serve as an intuitive understanding for the wide applicability of this pairing mechanism among various quasi-one- and two-dimensional systems. Since we neglect intersite correction by the DMFT part of the self-energy, the role of the local vertex corrections in the strong coupling regime, where the Mott transition and related phenomena take place, remains an important problem for the future studies. To this end, since non-local vertex corrections have to be considered, methods such as a method combining the FLEX approximation and the cluster extensions of DMFT would be promising.

References

1. Kuroki K, Higashida T, Arita R (2005) Phys Rev B 72:212509
2. Hirschfeld PJ, Korshunov MM, Mazin II (2011) Rep Progress Phys 74:124508
3. Miao H et al (2015) Nat Commun 6:6056
4. Wang F et al (2011) EPL (Europhys Lett) 93:57003
5. Bang Y (2014) New J Phys 16:023029
6. Chen X, Maiti S, Linscheid A, Hirschfeld PJ (2015) Phys Rev B 92:224514
7. Bang Y (2016) New J Phys 18:113054
8. Mishra V, Scalapino DJ, Maier TA (2016) Sci Rep 6:32078
9. Nakata M, Ogura D, Usui H, Kuroki K (2017) Phys Rev B 95:214509
10. Matsumoto K, Ogura D, Kuroki K (2018) Phys Rev B 97:014516
11. Matsumoto K, Ogura D, Kuroki K (unpublished)
12. Kobayashi K, Okumura M, Yamada S, Machida M, Aoki H (2016) Phys Rev B 94:214501
13. Maier TA, Scalapino DJ (2011) Phys Rev B 84:180513
14. Kuroki K, Kimura T, Arita R (2002) Phys Rev B 66:184508
15. Shinaoka H, Gull E, Werner P (2017) Comput Phys Commun 215:128
16. Gaenko A et al (2017) Comput Phys Commun 213:235
17. Bauer B et al (2011) J Stat Mech: Theory Exp 2011:P05001
18. Bergeron D, Tremblay A-MS (2016) Phys Rev E 94:023303
19. Mermin ND, Wagner H (1966) Phys Rev Lett 17:1133
20. Arita R, Kuroki K, Aoki H (2000) J Phys Soc Jpn 69:1181
21. Kitatani M, Tsuji N, Aoki H (2015) Phys Rev B 92:085104
22. Gukelberger J, Huang L, Werner P (2015) Phys Rev B 91:235114

Chapter 5
Possible High-T_c Superconductivity in "Hidden Ladder" Materials

Abstract In this chapter, we introduce a concept of "hidden ladder" for the bilayer Ruddlesden-Popper type compounds as a way to realize a ladder-like electronic structure: While the bilayer structure seems to have nothing to do with ladders, d_{xz} (d_{yz}) orbitals in the relevant t_{2g} sector of the transition metal form a two-leg ladder along x (y), since the d_{xz} (d_{yz}) electrons primarily hop in the leg (x, y) direction along with the rung (z) direction. This concept leads us to propose that $Sr_3Mo_2O_7$ and $Sr_3Cr_2O_7$ are candidates for the hidden-ladder material with the Fermi energy in the vicinity of the narrow band edge, which is suggested to work in favor of superconductivity by the previous study on the two-leg Hubbard ladder. Based on the first-principles band calculation and the analysis of the Eliashberg equation within the FLEX approximation, we propose a possible occurrence of high temperature superconductivity in these *non-copper* materials arising from the coexistence of a wide band and an "incipient" narrow band introduced in Chap. 2.

Keywords Hidden ladder · Incipient narrow band · Ruddlesden–Popper type materials

5.1 Motivation

In the preceding chapter, we have seen that the FLEX approximation is at least qualitatively valid even when the vertex corrections taken into account by DMFT is considered. Hence, a realization of high-T_c superconductivity in ladder type cuprates, proposed in Ref. [1] may indeed be possible, if a large amount of electron doping is attained. As mentioned earlier, however, such a large amount of doping is unrealistic for the cuprate ladders.

Therefore, in the present chapter, we intend to propose materials where the electronic structure is similar to the ladder cuprates, and still not so much carrier doping is required to realize the narrow incipient band situation favorable for high-T_c superconductivity. Here we adopt just FLEX, not FLEX+DMFT, since the electronic structure of the proposed materials is too complicated. Still, we believe that the conclusion is at least qualitatively valid on the basis of the analysis performed in Chap. 4.

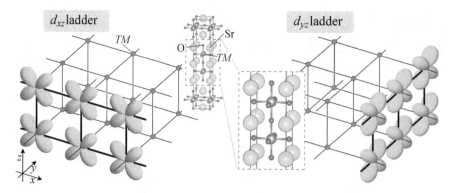

Fig. 5.1 Schematics of "hidden ladders" composed of the d_{xz} (left panel) and d_{yz} (right) orbitals in the bilayer Ruddlesden-Popper compounds $Sr_3TM_2O_7$ (*TM* indicates a transition metal). The middle panel depicts the crystal structure

Let us start with the Ruddlesden-Popper-type layered perovskites with two layers in a unit cell, whose crystal structure is depicted in the middle panel of Fig. 5.1. These bilayer compounds may first seem to have nothing to do with ladders. However, if we consider a case in which the t_{2g} orbitals are relevant (i.e., forming the bands crossing E_F), an electron in, say, the d_{xz} orbital selectively hops in the x direction as well as in the z direction normal to the bilayer. This means that the d_{xz} orbitals form a ladder with the x and z directions being the leg and rung directions, respectively. Similarly, d_{yz} orbitals form a ladder in the y and z directions, as schematically shown in Fig. 5.1. Hence ladder-like electronic structures are, in fact, *hidden* in the apparently bilayer Ruddlesden-Popper materials. To be precise, the d_{xz} and d_{yz} orbitals are weakly hybridized with each other, and also with the d_{xy} orbital, so that the quasi-one dimensional ladder character is slightly degraded, but they should still essentially be ladder bands.

5.2 Formulation

5.2.1 Intuitive Idea

As mentioned above, the band filling is a key factor if we want to make the narrow band incipient in the ladder-like electronic structure to realize a favorable condition for high-T_c superconductivity. According to the previous study, an electron doping ~30% is required for the ladder-type cuprate, where the $d_{x^2-y^2}$ (i.e. an e_g) orbital is relevant. In the case when the d_{xz} and d_{yz} (i.e. t_{2g}) orbitals are relevant, the ideal situation should be ~30% *hole* doping from half filling since the roles of electrons and holes are exchanged from those in the $d_{x^2-y^2}$ orbital case as can be seen by considering the sign of atomic orbitals contributing to the second neighbor hopping integral. Assuming the crystal field splitting between d_{xy} and $d_{xz/yz}$ is not so large, this situation corresponds to 1/3 filling, namely, two electrons in the three t_{2g} orbitals. If we consider non-copper materials of the form $Sr_3TM_2O_7$ (*TM*: transition metal), the valence of the transition metal should be +4. In order to have a d^2 electron configuration, we can take, instead of hole doping, stoichiometries $Sr_3Mo_2O_7$ (with a 4d transition metal Mo) and $Sr_3Cr_2O_7$ (with a 3d transition metal Cr) for possible candidates for the hidden ladder materials with the right electron configuration.

5.2.2 Band Calculation and Many-body Analysis

We start with a first principles band calculation for these materials with the WIEN2K package [2], assuming the ideal case of no long-range orders and adopting the experimental crystal structure with $I4/mmm$ space group in Refs. [3, 4]. Since the experimental structure for $Sr_3Mo_2O_7$ is taken from the experiment for a zinc-replaced compound $Sr_3Mo_{1.5}Zn_{0.5}O_{7-\delta}$, we also performed the structural optimization calculation for $Sr_3Mo_2O_7$ using the VASP package [5, 6]. Structural parameters of these two structures for $Sr_3Mo_2O_7$ are listed in Table 5.1. Overall structures are consistent with each other within about 1% of accuracy. We note that in the optimized structure, the distance between two Mo-O layers becomes shorter than that in the experimental structure, which may lead to a somewhat stronger bilayer coupling. In the band calculation, we employ the PBE-GGA functional with 1000 k-meshes and $RK_{max} = 7$.

To proceed to many-body analysis, we construct a three-dimensional tight-binding model comprising six orbitals (three t_{2g} orbitals of Mo in each layer) for $Sr_3Mo_2O_7$ exploiting the maximally-localized Wannier orbitals with the WANNIER90 code [7] and the WIEN2WANNIER interface [8].

To take into account the electron correlation effect beyond the GGA level, we consider multi-orbital Hubbard type interactions on each atomic site represented by Eq. (3.12) on top of the tight-binding model and apply the multi-orbital FLEX approximation. We set the on-site interactions as $U = 2.5$ eV (or $U = 3.0$ eV) and

Table 5.1 Structural parameters for $Sr_3Mo_2O_7$ for the experimental and optimized structure. Mo-apical O(outer/inner) means the distance between Mo and O located at outside/between the bilayer. Mo-in-plane O means the distance between Mo and in-plane O

	$Sr_3Mo_2O_7$ (experimental)	$Sr_3Mo_2O_7$ (optimized)
a-axis (Å)	3.96	4.01
c-axis (Å)	20.54	20.64
Mo-apical O (outer) (Å)	2.02	2.00
Mo-apical O (inner) (Å)	2.02	2.06
Mo-in-plane O (Å)	1.98	2.01

$J = U/10$ along with $U' = U - 2J$ (for preserving the orbital rotational invariance). These are typical values among the estimations by the constrained RPA [9–11] or the constrained LDA [12] method for the Hubbard-type models for the d-orbitals of transition metal. The temperature is fixed at $k_B T = 0.01$ eV and we take $32 \times 32 \times 4$ k-meshes and 2048 Matsubara frequencies.

As mentioned in the beginning of this chapter, since calculations involving the strong local correlation effects for the six-orbital models are computationally too expensive, we employ the FLEX approximation in the present chapter. Still, we expect the conclusion would qualitatively be unaffected since a pairing mechanism exploiting an incipient band discussed in the previous studies is still valid even with the local vertex corrections as we discussed in Chap. 4.

5.3 Results

5.3.1 Band Structure and Effective Model

Firstly, we show the band structures of $Sr_3Mo_2O_7$ and $Sr_3Cr_2O_7$ obtained from the first-principles band calculation the panels (a–f) of Fig. 5.2. The thickness of the lines in the left(right) panels represents the weight of the $d_{xy}(d_{xz,yz})$ orbital character. This result indeed shows that the d_{xz} and d_{yz} orbitals form two sets of quasi-one-dimensional narrow (antibonding) and wide (bonding) bands in both materials, as one can see from the small dispersion along P-N and N-Γ. We can also see that the hybridization between the d_{xy} and $d_{xz,yz}$ orbital is substantially small. Another important observation is that the band structure of $Sr_3Mo_2O_7$ is very similar to that of $Sr_3Cr_2O_7$. Therefore, in the many-body analysis, we focus on $Sr_3Mo_2O_7$ and expect similar feature in $Sr_3Cr_2O_7$.

In Fig. 5.3, the density of states of $Sr_3Mo_2O_7$ using the experimental and optimized structure as well as that of $Sr_3Cr_2O_7$. We also show the projected density of states for *TM* along with the d_{xy} and $d_{xz,yz}$ orbital of *TM*. It can be seen that the Fermi level sits just below the large density of states which originate from the $d_{xz,yz}$ orbital.

In order to compare these results to a ladder cuprate, we also performed a band calculation for the well-known two-leg ladder cuprate $SrCu_2O_3$ using the crystal structure in Ref. [13, 14] and display the result in Fig. 5.4(e,f) along two paths in the conventional Brillouin zone so as to make the relation to the Ruddlesden–Popper compounds clear. Note that the sign of the energy is reversed in order to facilitate comparison between the t_{2g} and e_g systems. Superimposing panels (a) and (b) in Fig. 5.4(e,f), we can see that the band structures of the $d_{xz,yz}$ orbital in the Ruddlesden–Popper compounds are quite similar to that of the ladder cuprate. However, an important difference lies in the position of the Fermi energy E_F. Namely, while E_F is located at very close to the bottom edge of the narrow bands in the Ruddlesden–Popper compounds, E_F lies in deep inside of the narrow band. In other words, the narrow bands in the Ruddlesden–Popper compounds can be regarded

5.3 Results

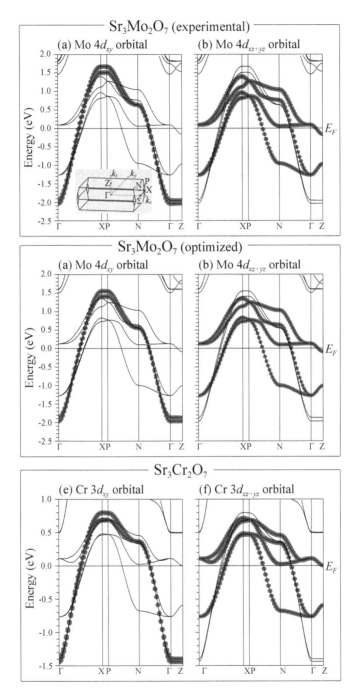

Fig. 5.2 Band structures of **a**, **b** $Sr_3Mo_2O_7$ for the experimental structure, **c**, **d** $Sr_3Mo_2O_7$ for the optimized structure, and **e**, **f** $Sr_3Cr_2O_7$ obtained from the first-principles calculation. The Brillouin zone is shown in the inset. The size of the blue circles in the left (right) panels depicts the weight of the $d_{xy}(d_{xz,yz})$ orbital character (Color figure online)

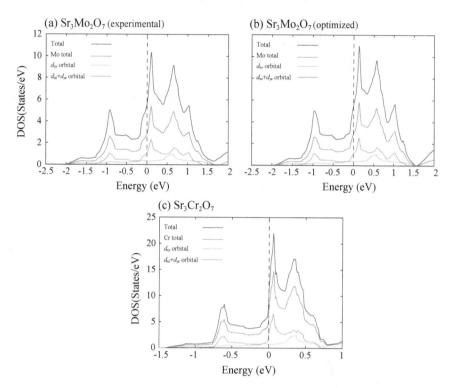

Fig. 5.3 Density of states of **a** $Sr_3Mo_2O_7$ for the experimental structure, **b** $Sr_3Mo_2O_7$ for the optimized structure, and **c** $Sr_3Cr_2O_7$ obtained from the first-principles calculation. Purple lines denote the total density of states along with the projected density of states for the *TM* atom, the *TM* d_{xy} orbital, and the *TM* d_{xz+yz} orbital by green, orange, and cyan lines, respectively. Dashed vertical lines represents the Fermi level

as incipient band *without* hole doping. These results confirm the above-mentioned intuitive expectation.

In Fig. 5.5, we superpose the band structures of the six-orbital tight-binding model for the $Sr_3Mo_2O_7$ constructed from the maximally-localized Wannier functions to that of the first-principles calculation. We can see that the fit is quite accurate for all the bands intersecting the Fermi level. To be precise, since hopping processes between Mo atoms are mainly mediated by oxygen p orbitals in the original material, the "t_{2g}" Wannier orbitals in the tight-binding model are interpreted as the orbitals in which the contribution of the p orbitals are effectively taken into account, and hence different from those in the original one. However, since the bands intersecting the Fermi level are well separated from the other bands, we still simply call them t_{2g} orbitals if there is no misunderstanding. We also list the hopping integrals for the d_{xz} orbital in the six-orbital model for $Sr_3Mo_2O_7$ and $Sr_3Cr_2O_7$ (for comparison) in Table 5.2, where t_x^{intra} denotes the intralayer nearest neighbor hopping towards the x direction (i.e. leg of a ladder), t_d^{inter} denotes the interlayer hopping within the unit cell

5.3 Results

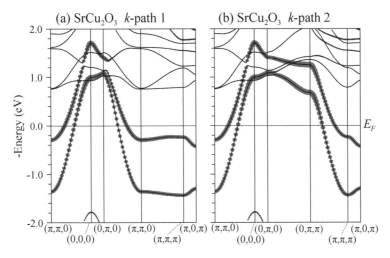

Fig. 5.4 The band structure of a ladder cuprate SrCu$_2$O$_3$ is displayed along two paths in the conventional Brillouin zone, where the thickness of the lines represents the $d_{x^2-y^2}$ orbital character. The point $(\pi, \pi, 0)$ corresponds to Γ point in Fig. 5.2 because the sign of the hopping is reversed. Note that panels (**a**) and (**b**) superimposed form a band structure quite similar to those in Fig. 5.2b, d, respectively, except for the position of the Fermi level

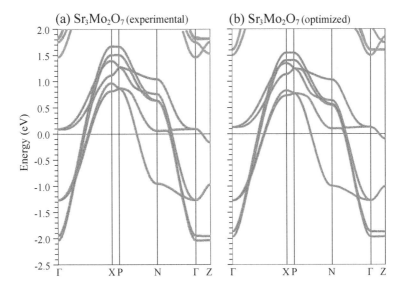

Fig. 5.5 Band structure (red lines) of the six-orbital tight-binding model for Sr$_3$Mo$_2$O$_7$ constructed from the maximally-localized Wannier orbitals. Gray lines represent the first-principles calculation

Table 5.2 Hopping integrals within the d_{xz} sector of the six-orbital tight-binding model for $Sr_3Mo_2O_7$ for the experimental structure, $Sr_3Mo_2O_7$ for the optimized structure, and $Sr_3Cr_2O_7$. Each column is described in the main text

	t_x^{intra} (eV)	t_d^{inter} (eV)	t_x^{inter} (eV)	t_y^{intra} (eV)
$Sr_3Mo_2O_7$ (experimental)	−0.376	−0.326	−0.093	−0.048
$Sr_3Mo_2O_7$ (optimized)	−0.357	−0.374	−0.089	−0.044
$Sr_3Cr_2O_7$	−0.204	−0.207	−0.071	−0.035

(i.e. rung of a ladder), t_x^{inter} denotes the interlayer nearest neighbor hopping towards the x direction, which controls the difference in the band width between the wide and narrow bands, and t_y^{intra} denotes the intralayer hopping towards the y direction, which controls two-dimensionality. These values also confirm that the d_{xz} orbital essentially forms a ladder toward the x (leg) and z (rung) directions. Apparently, in the Mo-compound, the band structure for the optimized lattice structure is quite similar to that for the experimental crystal structure, it is, however, important that t_d^{inter} for the optimized structure is rather large compared to that for the experimental structure, which leads to a larger bonding-antibonding splitting and hence may affect the electron correlation effects.

5.3.2 Many-body Analysis

Using the above six-orbital model, we now study the possibility of superconductivity of $Sr_3Mo_2O_7$. Figure 5.6 displays the FLEX result for the eigenvalue λ of the Eliashberg equation against the band filling. We also show the filling dependence of the Stoner factor α_S which is defined as the maximum eigenvalue $\hat{\chi}^{irr}(q)\hat{S}$ and we use it to measure how close the system is to magnetic instabilities. Here the filling n is defined as the average number of electrons per unit cell, so that $n = 4$ corresponds to the stoichiometric d^2 electron configuration. Comparing the result for the experimental structure and that for the optimized structure, whereas we can see that superconductivity is optimized around the stoichiometric band filling for both structures, the optimal filling is somewhat different, $n \simeq 3.8$ for the experimental structure and $n \simeq 4.2$ for the optimized structure. This would be originating from the difference in the position of the narrow band edge. In the optimized structure, the narrow band edge slightly shifts to higher energy in comparison to the case of the experimental structure, which is due to the enhancement of t_d^{inter}. This would make the optimal band filling larger as in the result. As for the choice of the interaction parameters, λ is qualitatively unchanged when we change $U = 2.5$–3.0 eV within a realistic range for the model of $Sr_3Mo_2O_7$ for the experimental structure, except that for $U = 2.5$ eV, (i) the maximum value of λ is somewhat enhanced from the case

5.3 Results

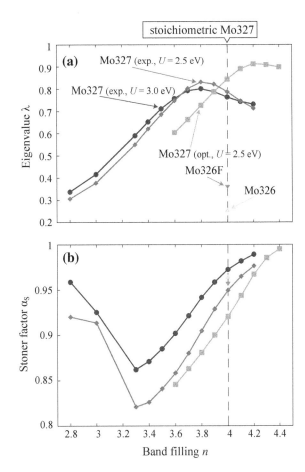

Fig. 5.6 FLEX result for the band filling dependence of **a** the eigenvalue λ of the linearized Eliashberg equation and **b** the Stoner factor for $Sr_3Mo_2O_7$ for $U = 2.5$ eV (red diamonds), $U = 3.0$ eV (purple diamonds) for the experimental structure and for $U = 2.5$ eV for the optimized structure. The vertical dashed line represents the stoichiometric point ($n = 4$). Also displayed are the results for the 326 compound $Sr_3Mo_2O_6$ (orange triangle) and a F-doped $Sr_3Mo_2O_6F$ (blue inverted triangle) at $n = 4$ (Color figure online)

of $U = 3$ eV, and (ii) a decrease of λ in the left side of the dome becomes slightly steeper. We can see that while the Stoner factor α_S monotonically decreases with reducing the band filling for $n \gtrsim 3.3$, further decreasing n, α_S becomes large again. As we discuss later, this enhancement is not relevant to the $d_{xz,yz}$ orbital and hence superconductivity which we are considering.

In any case, we can see that λ attains a maximum $\simeq 0.8$ near the stoichiometric band filling of $n = 4$ in all of cases, which can be regarded to signal a high-T_c in this material, since the value of λ is similar to that of the cuprate superconductor $HgBa_2CuO_4$ having $T_c \simeq 90$ K as obtained with FLEX for systematic models that qualitatively explains the material dependence of T_c among the cuprates [15]. We might expect even larger λ (higher T_c) if we could apply FLEX+DMFT to the present system, considering the tendency of the difference between FLEX and FLEX+DMFT observed in Chap. 4. To be precise, λ in Fig. 5.6 is peaked at $n \simeq 3.8$ or $\simeq 4.2$, but decreases as we move away from that point, which can be interpreted as follows: for a too small band filling, superconductivity is suppressed because the narrow band lies too far from the Fermi level. When the band filling is too large, on the other hand, the Fermi level lies deep inside the narrow band, so that the system is put close to a magnetic order as can be seen from the Stoner factor, suppressing λ. Since the results are qualitatively unchanged in any cases, hereafter we will discuss the details on the model for $Sr_3Mo_2O_7$ for the experimental structure with $U = 2.5$ eV.

In order to explore which pair structure is obtained from the linearized Eliashberg equation, first, we plot the obtained anomalous self-energy $\Delta(k)$ at the lowest Matsubara frequency in the orbital representation in Fig. 5.7. Here the orbital indices $\{0, 1, 2, 3, 4, 5\}$ denote the $\{d_{xy}^1, d_{xz}^1, d_{yz}^1, d_{xy}^2, d_{xz}^2, d_{yz}^2\}$ orbitals, with the superscript 1 or 2 indicating the layer index. For a small filling ($n = 3.3$), both the intra-layer ($i = j$) and inter-layer components ($i \neq j$) of $\Delta_{ij}(k, i\pi k_B T)$ in Fig. 5.7 are small compared to those of the other fillings. Another important feature is that the intra-layer components do not change the sign, namely intra-layer repulsive interactions are always harmful to superconductivity corresponding to this $\Delta(k, i\pi k_B T)$. For a large filling ($n = 4.2$), both the intra-layer and inter-layer components have strong momentum dependence and have comparable magnitude with each other. For the optimal filling, the inter-layer components are dominant and the intra-layer components reverse their sign. For all the fillings, we can see that the d_{xy} orbital components

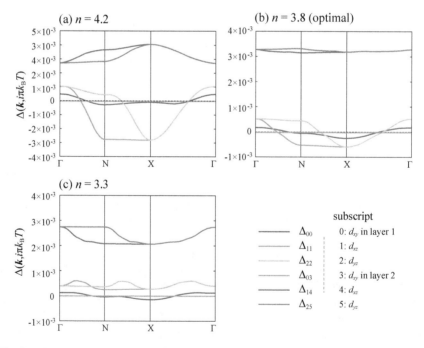

Fig. 5.7 Obtained anomalous self-energy for **a** $n = 4.2$, **b** $n = 3.8$ (optimial filling for superconductivity), and **c** $n = 4.2$. The subscripts $\{0, 1, 2, 3, 4, 5\}$ denote the $\{d_{xy}^1, d_{xz}^1, d_{yz}^1, d_{xy}^2, d_{xz}^2, d_{yz}^2\}$ orbitals, with the superscript 1 or 2 indicating the layer index

5.3 Results

do not have large contribution, namely we can say that main players of superconductivity in this system are electrons in the d_{xz} and d_{yz} orbitals as we expected.

Next, let us study how the gap structure looks like in the band representation, which would correspond to the gap function observed in experiments. To convert $\Delta(\bm{k}, i\pi k_\mathrm{B} T)$ to the band representation, we perform the unitary transformation which diagonalizes the tight-binding Hamiltonian. However, since each band is labeled by ascending order from the small energy eigenvalue at each $\bm{k}$ in the numerical calculation, the momentum dependence of the unitary transformation can be inconsistent with the character of the eigenvector. This makes the band representation of $\Delta(\bm{k}, i\pi k_\mathrm{B} T)$ difficult to interpret. To circumvent this difficulty, we take linear combinations of the intra-layer and inter-layer components as $\Delta_{i,i} + \Delta_{i,i+3}$ and $\Delta_{i,i} - \Delta_{i,i+3}$ for $i = 0, 1, 2$, which correspond to the bonding and anti-bonding orbital components. In Fig. 5.8, we show the band representation of $\Delta(\bm{k}, i\pi k_\mathrm{B} T)$ and these linear combinations. As we can see in Fig. 5.8, the above linear combinations well reproduce the band representation of $\Delta(\bm{k}, i\pi k_\mathrm{B} T)$, namely the Cooper pair considered here is mainly formed across the two planes (at least around the

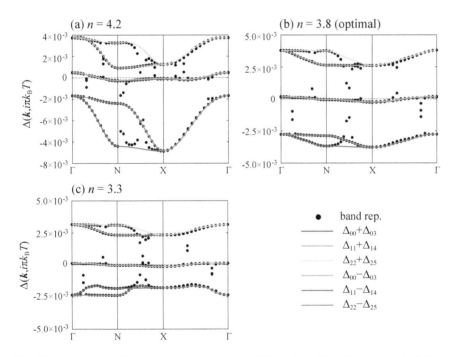

Fig. 5.8 Anomalous self-energy for **a** $n = 4.2$, **b** $n = 3.8$ (optimal filling for superconductivity), and **c** $n = 4.2$ in the band representation (black dots) obtained with the unitary transformation which diagonalizes the six-orbital tight-binding Hamiltonian. For intuitive understanding, linear combinations of the intra- and inter-layer component for the $d_{xy}/d_{xz}/d_{yz}$ orbital are also shown. Purple, green, and cyan lines correspond to the bonding orbitals comprising the d_{xy}, d_{xz}, and d_{yz} orbital, respectively, and, orange, yellow, and blue lines to the anti-bonding orbitals

optimal filling) so that the gap function for the bonding and anti-bonding orbitals have the opposite sign. This indeed implies high-T_c superconductivity expected in this system is essentially the same as that proposed in Ref. [1] in the Hubbard model on a two-leg ladder.

In the left panels of Fig. 5.9, we show the renormalized density of states ($A(\omega)/2\pi$, $A(\omega)$: spectral function) obtained with the Padé approximation. While the projected density of states of the d_{xy} orbital barely depends on the filling, that of the $d_{xz,yz}$ orbitals strongly depends on the filling. This implies the inter-orbital interactions between the d_{xy} and $d_{xz,yz}$ orbitals do not play an important role in the filling dependent electron correlation effects such as the dependence of λ. In order to obtain an intuitive understanding on the reason for the enhancement of superconductivity, let us take a closer look at this renormalized electronic structure. As we have seen above, the Cooper pair considered here is mainly formed across the two planes so that the gap function for the bonding and anti-bonding orbitals have the opposite sign. So we decompose the local Green's function into the contribution from the bonding/anti-bonding orbital by the unitary transformation U defined by

$$U = U_{\text{orbital}} \otimes U_{\text{layer}}$$

$$= \begin{pmatrix} 1 & 0 & 0 \\ 0 & 1 & 0 \\ 0 & 0 & 1 \end{pmatrix} \otimes \frac{1}{\sqrt{2}} \begin{pmatrix} 1 & -1 \\ 1 & 1 \end{pmatrix}, \tag{5.1}$$

and perform the analytic continuation with the Padé approximation. The decomposed density of states for some typical band fillings are shown in the right panels of Fig. 5.9. For a too small band filling ($n = 3.3$), the anti-bonding orbital with a narrow band width does not have weight at the zero energy (i.e. the Fermi level), whereas the bonding orbital with a wide band width has sharp peaks. This means that an electron in the bonding orbital is not strongly renormalized. Moreover, this also suggests that the interband pairing interaction between the wide band mainly comprising the bonding orbital and the narrow band mainly comprising the anti-bonding orbital is weak, and hence superconductivity is degraded. On the other hand, for a too large band filling ($n = 4.2$), we can see that both the bands have considerable weight at the Fermi level, and the wide band is strongly renormalized since there are no sharp structures. This leads to the suppression of λ. Hence we conclude that the superconductivity is enhanced in an intermediate filling region, around the stoichiometric point ($n = 4$).

We also show similar plots for $n = 2.8$, a filling with large Stoner factor, in Fig. 5.10. It can be seen that the renormalized density of states of the $d_{xz,yz}$ orbital is not so strongly renormalized. This means that the enhancement of the Stoner factor around $n = 2.8, 3.0$ is not due to the $d_{xz,yz}$ orbitals. Namely that enhancement is not relevant to the Cooper pair formation which we are considering here.

If we turn from the Mo compound to the $3d$ compound, $Sr_3Cr_2O_7$, we may also expect high-T_c superconductivity, since its band structure is similar to that of $Sr_3Mo_2O_7$ as seen in Fig. 5.2.

5.3 Results

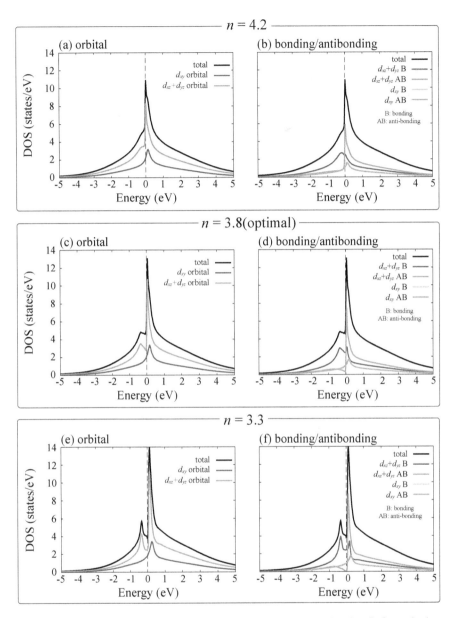

Fig. 5.9 Renormalized density of states obtained with the Padé approximation. Left panels show the projected density of states of the $d_{xz,yz}$ (purple lines) and d_{xy} orbitals (green) along with the total density of states (black). Right panels show the projected density of states of $d_{xz,yz}/d_{xy}$ decomposed into the bonding (purple/cyan) and anti-bonding orbitals (green/orange). The vertical dashed line represents the chemical potential

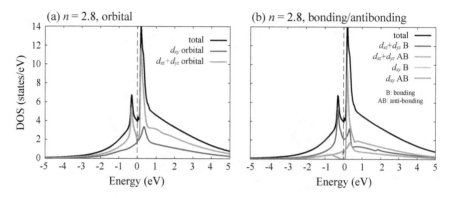

Fig. 5.10 Renormalized density of states obtained with the Padé approximation for $n = 2.8$. Panels are similar to Fig. 5.9. The vertical dashed line represents the chemical potential

5.4 Discussion

5.4.1 Relation to Experimental Observations in Actual Materials

So far, we have discussed a possible occurrence of a high-T_c superconductivity in $Sr_3Mo_2O_7$ and $Sr_3Cr_2O_7$. However, the materials, $Sr_3Mo_2O_7$ and $Sr_3Cr_2O_7$, have in fact been synthesized in the past [16–18]. $Sr_3Cr_2O_7$ is reported as being an antiferromagnetic insulator, while $Sr_3Mo_2O_7$ a Pauli-paramagnetic metal, where superconductivity has not been observed so far in these materials. Let us then explore the reasons for that. $Sr_3Cr_2O_7$ has two d electrons within the three t_{2g} orbitals, so that we have a possibility of orbital ordering and the Mott transition to put the system in an insulating state. Actually, a recent experiment reports that an orbital ordered state with lattice distortion is observed in $Sr_3Cr_2O_7$ [19]. This kind of instabilities can not be captured in the present formalism since it requires to handle strong electron correlations involving multiple orbitals. In order to realize superconductivity, such orders have to be suppressed by, e.g., applying pressure or doping carriers. If we apply a physical pressure, for instance, the system will become more weakly correlated due to an increased band width. Applying chemical pressure with isovalent elemental substitution such as Ca doping to Sr sites can also exert a similar effect. As for the carrier concentration, carrier doping in the $d_{xz/yz}$ orbitals can suppress the orbital ordering and hence the insulating state, but chemical substitution may be difficult to achieve. One way to attain this is to dope carriers to the $d_{xz/yz}$ orbitals effectively by applying uniaxial pressure/strain along the c-axis (in a single crystal) to control the level offset between the d_{xy} and $d_{xz/yz}$ orbitals. Another possible way for carrier doping is to utilize field-effect transistor techniques, which enables fine tuning of the carrier concentration.

In the case of $Sr_3Mo_2O_7$, the situation might be more complicated since it is known to be a Pauli paramagnetic metal which implies that the system is, in contrast to

5.4 Discussion

$Sr_3Cr_2O_7$, a typical weakly-correlated one. Regarding this, we speculate that oxygen deficiencies, especially at the sites between adjacent MoO planes, may be detrimental to superconductivity. Since the Cooper pair considered here is formed across the two planes (i.e., on each rung of a ladder) in real space, the loss of these oxygens would hinder the superconductivity. Indeed, Ref. [20] reports that in $La_3Ni_2O_7$ having the same bilayer Ruddlesden-Popper structure as $Sr_3Mo_2O_7$, oxygen vacancies are located solely at the sites between adjacent NiO layers [21]. As for $Sr_3Mo_2O_7$, there is in addition some difficulty in the synthesis of samples with the designated stoichiometric composition [22]. In Ref. [17], there remains ambiguity in the precise chemical composition, since the samples are fabricated with an oxygen buffer, Ti_2O_3. In high-pressure synthesis methods, the samples tend to be obtained with oxygen contents ($\simeq O_{6.3}$) somewhat reduced from the nominal one [22]. Thus the oxygen deficiencies at the sites between adjacent MoO planes are very likely to be present in the samples fabricated so far.

In order to demonstrate that deficiencies of the oxygen atoms connecting the two planes are fatal for superconductivity, we have actually performed a calculation for two hypothetical materials: a "326" compound $Sr_3Mo_2O_6$ and a F-doped $Sr_3Mo_2O_6F$. The crystal structure of $Sr_3Mo_2O_6$, where the oxygens connecting two MoO layers are totally missing, is determined with the structural optimization calculation using the VASP package [5, 6]. Fluorine-replaced $Sr_3Mo_2O_6F$, on the other hand, mimics $Sr_3Mo_2O_{6.5}$, of which first-principles calculations necessitate to take a large unit cell, which makes a many-body calculation too expensive. We employ the structure average between optimized structures of $Sr_3Mo_2O_7$ and $Sr_3Mo_2O_6$ for this hypothetical material. The obtained band structures and the hopping integrals in the six-orbital model constructed by the maximally-localized Wannier functions are shown in Fig. 5.11 and Table 5.3. By comparing these to the case of $Sr_3Mo_2O_7$, one can indeed see that the ladder-like electronic structure is strongly degraded by oxygen deficiencies at the site connecting two planes: hopping integrals connecting adjacent MoO planes (t_d^{inter} and t_x^{inter}) are strongly suppressed by oxygen deficiencies, namely, the electronic structure in the $d_{xz/yz}$ orbitals becomes chain-like. In order to see how this degraded electronic structure affects superconductivity, we perform a FLEX calculation for the six-orbital model for each material as in the case of $Sr_3Mo_2O_7$. $U = 2.5$ eV is adopted here. The obtained eigenvalues superposed in Fig. 5.6a indeed show that the oxygen deficiencies drastically suppress superconductivity. To show this, we have fixed the filling at $n = 4$ to single out the effect of the degraded ladder-like structure, oxygen deficiencies should introduce some electron doping, which makes the Fermi energy deeper into the narrow band. If we compare these results for the oxygen defected systems to that for $Sr_3Mo_2O_7$ with the experimental structure, electron doping may further degrade superconductivity since the optimal band filling is in the hole-doped region. On the other hand, if we compare to $Sr_3Mo_2O_7$ with the optimized structure, electron doping might lead to an enhancement of λ in a certain extent, but still we would expect that enhancement is not strong enough to have comparable λ to that for $Sr_3Mo_2O_7$. Thus the direction suggested from these arguments is to synthesize the material with as little oxygen deficiencies as possible.

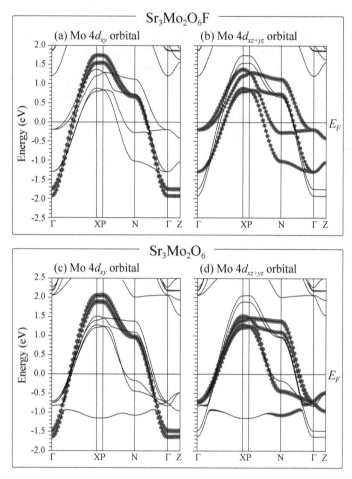

Fig. 5.11 Band structures of **a**, **b** $Sr_3Mo_2O_6F$, **c**, **d** $Sr_3Mo_2O_6$. The size of the blue circles in the left (right) panels depicts the weight of the d_{xy} ($d_{xz,yz}$) orbital character (Color figure online)

Table 5.3 Hopping integrals within the d_{xz} sector of the six-orbital tight-binding model for hypothetical materials $Sr_3Mo_2O_6F$ and $Sr_3Mo_2O_6$

	t_x^{intra} (eV)	t_d^{inter} (eV)	t_x^{inter} (eV)	t_y^{intra} (eV)
$Sr_3Mo_2O_6F$	−0.421	−0.290	−0.052	−0.032
$Sr_3Mo_2O_6$	−0.514	−0.160	−0.028	−0.023

5.4.2 Other Candidates for Hidden Ladder

While the present study focuses on bilayer Ruddlesden-Popper materials, the present idea can be extended to some other bilayer materials, such as fluorine-intercalated compounds $La_2SrCr_2O_7F_2$ and $La_2SrMo_2O_7F_2$. In these compounds, fluorine atoms are intercalated between the bilayer structures that are the same as those in $Sr_3Cr_2O_7$ and $Sr_3Mo_2O_7$, so that the band structure should be quite similar. Also in these compounds, the valence of the transition metal should be +4, namely a d^2 electron configuration. In order to confirm this expectation, we have performed a first-principles band calculation for the ideal case of no long-range orders or lattice distortion. First we determined the crystal structure of $La_2SrCr_2O_7F_2$ and $La_2SrMo_2O_7F_2$ with the VASP package. The virtual crystal approximation(VCA) is adopted to take account of the effect of the substitution of La for Sr. The band structures obtained with the optimized lattice structures are shown in Fig. 5.12, which are indeed quite similar to those of $Sr_3TM_2O_7$ (TM = Mo, Cr) except that the three dimensionality is somewhat reduced. It is important to note that the Fermi level lies in the vicinity of the narrow band edge. Given the band structure similar to those of the Ruddlesden-Poppers 327 systems, high-T_c superconductivity is again expected in these systems under certain conditions. In fact, $La_2SrCr_2O_7F_2$ has been synthesized [23], but turns out to be an antiferromagnetic insulator with tilted CrO octahedra. As in $Sr_3Cr_2O_7$, suppression of the magnetic order (and possibly the lattice distortion) by carrier doping, etc. should be necessary for superconductivity.

We can further extend the present concept for the double-layer compounds to *triple*-layer cases, e.g., $A_4TM_3O_{10}$, where we can envisage that three-leg-ladder-like electronic structures are hidden. Namely, if one adds the third layer in Fig. 5.1

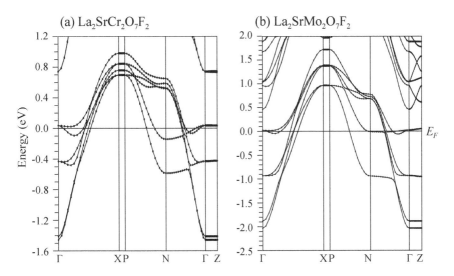

Fig. 5.12 Band structures of fluorine-intercalated compounds **a** $La_2SrCr_2O_7F_2$ and **b** $La_2SrMo_2O_7F_2$

to consider the d_{xz} and d_{yz} orbitals in a similar manner, one can find a pair of three-leg ladders as schematically depicted in Fig. 5.13. Note that, unlike the case of the bilayer compounds, since a favorable carrier concentration is unclear in this case, one has to look for the optimal stoichiometry. In order to show that the above expectation is valid, we have performed a band calculation [24] for two compounds, $Sr_4Cr_3O_{10}$ and $La_4Cr_3O_{10}$. Whereas $Sr_4Cr_3O_{10}$ has actually been synthesized in the past [16], $La_4Cr_3O_{10}$ has not been synthesized so far. To make a fair comparison between them, we have theoretically determined the crystal structures of them with the structure optimization. The resulting band structures are displayed in Fig. 5.14. For comparison, we have also performed a band calculation for a three-leg ladder cuprate $Sr_2Cu_3O_5$, with the lattice structure given in Ref. [25]. The result is shown in the bottom panels of Fig. 5.14 with the reversed sign of the energy to facilitate comparison as in the two-leg ladder case. Since there are two inequivalent [inner- and outer-layer(chain)] Cr(Cu) sites, we show the orbital characters of $t_{2g}(3d_{x^2-y^2})$ orbitals for each of them. The band structure of the d_{xz} and d_{yz} orbitals is indeed seen to be similar to that of the three-leg ladder. As explained in Chap. 2, theoretically, it has been known that three-leg ladders should also exhibit superconductivity [26–29] as in the two-leg case. According to Ref. [30], superconductivity is strongly enhanced when the chemical potential is in the vicinity of the edge of the narrow band even in the

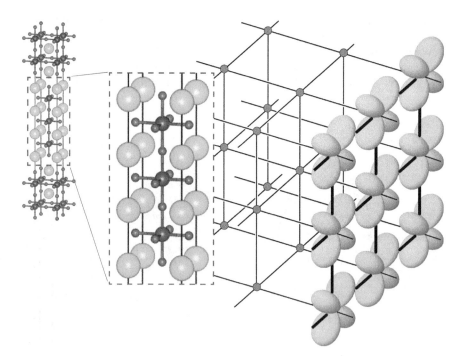

Fig. 5.13 A schematic hidden ladder in the triple-layer Ruddlesden-Popper materials, here displayed for d_{yz} orbitals

5.4 Discussion

three-leg Hubbard ladder model. The narrow band edge for these materials is depicted by red dashed lines in Fig. 5.14, but the three-leg ladder cuprate $Sr_2Cu_3O_5$ itself is well-known for being difficult to dope carriers. Observing the band calculation results of the triple-layer materials, we may expect high-T_c superconductivity if (La, Sr)$_4$Cr$_3$O$_{10}$ can be synthesized with a certain content ratio of La and Sr, presumably Sr-rich.

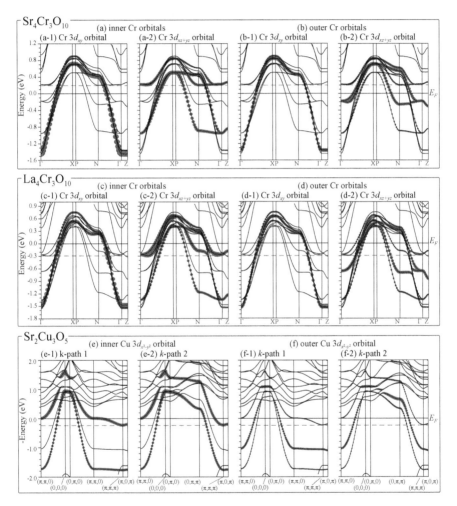

Fig. 5.14 Band structures of triple-layer compounds **a, b** $Sr_4Cr_3O_{10}$ and **c, d** $La_4Cr_3O_{10}$. The thickness of the lines in the left(right) panels depicts the weight of the $d_{xy}(d_{xz,yz})$ orbital character. **e, f** For comparison, the band structure of a three-leg ladder cuprate $Sr_2Cu_3O_5$ is shown along two paths in the conventional Brillouin zone, where the thickness of the lines represents the $d_{x^2-y^2}$ character. Note that the sign of the energy is reversed to facilitate comparison between the t_{2g} (present) and e_g (cuprate) systems. As in the two-layer case, panels (e–1) and (e–2) [(f–1) and (f–2)] superimposed form a band structure quite similar to those in (a–2) and (c–2) [(b–2) and (d–2)]. Red dashed lines indicate the edge of the narrow bands (Color figure online)

5.5 Summary

To summarize, we have introduced the concept of the "hidden ladder" electronic structure in the Ruddlesden-Popper layered perovskites, where anisotropic d orbitals of the transition metal give rise to inherent ladder-like electronic structures. We have proposed actual candidates in $Sr_3Mo_2O_7$ and $Sr_3Cr_2O_7$ from first-principles band calculations. We have then pointed out that coexisting narrow and wide bands arising from the ladder can host high-temperature superconductivity due to the strong pairing interaction originating from the virtual pair-hopping processes between the wide and incipient narrow bands, which is supported by a FLEX calculation for $Sr_3Mo_2O_7$, where no or only slight carrier doping is required. The idea can be further extended to fluorine intercalated and triple-layer compounds. To realize superconductivity, we note that the possibility of Mott transition as well as the occurrence of other orders such as orbital ordering has to be considered, details of which will be an important future work.

References

1. Kuroki K, Higashida T, Arita R (2005) Phys Rev B 72:212509
2. Blaha P, Schwarz K, Madsen G, Kvasnicka D, Luitz J (2001) An augmented plane wave+ local orbitals program for calculating crystal properties
3. Steiner U, Reichelt W, Schmidt M, Schnelle W (2004) Zeitschrift für anorganische und allgemeine Chemie 630:649
4. Castillo-Martínez E, Alario-Franco MÁ (2007) Solid State Sci 9:564
5. Kresse G, Furthmüller J (1996) Phys Rev B 54:11169
6. Kresse G, Joubert D (1999) Phys Rev B 59:1758
7. Mostofi AA et al (2008) Comput Phys Commun 178:685
8. Kuneš J et al (2010) Comput Phys Commun 181:1888
9. Vaugier L, Jiang H, Biermann S (2012) Phys Rev B 86:165105
10. Mravlje J et al (2011) Phys Rev Lett 106:096401
11. Jang SW et al (2016) Sci Rep 6:33397
12. Pchelkina ZV et al (2007) Phys Rev B 75:035122
13. Sparta K, Löffert A, Gross C, Aßmus W, Roth G (2006) Zeitschrift für Kristallographie-Crystalline Materials 221:782
14. Müller TFA, Anisimov V, Rice TM, Dasgupta I, Saha-Dasgupta T (1998) Phys Rev B 57:R12655. in which band calculation for $SrCu_2O_3$ has been performed
15. Sakakibara H, Usui H, Kuroki K, Arita R, Aoki H (2010) Phys Rev Lett 105:057003
16. Kafalas J, Longo J (1972) J Solid State Chem 4:55
17. Kouno S et al (2007) J Phys Soc Jpn 76:094706
18. Hosono H et al (2015) Sci Technol Adv Mater 16:033503
19. Jeanneau J et al (2017) Phys Rev Lett 118:207207
20. Poltavets VV et al (2006) J Am Chem Soc 128:9050
21. Pardo V, Pickett WE (2011) Phys Rev B 83:245128. in which a band structure of $La_3Ni_2O_7$ is obtained for various levels of oxygen deficiency as well as lattice structure distortions
22. Eisaki H, Private communications
23. Zhang R et al (2016) Inorg Chem 55:3169
24. These calculations for the trilayer compounds have been done in collaboration with Y. Hirabuki
25. Kazakov S et al (1997) Phys C: Supercond 276:139

References

26. Arrigoni E (1996) Phys Lett A 215:91
27. Schulz HJ (1996) Phys Rev B 53:R2959
28. Kimura T, Kuroki K, Aoki H (1996) Phys Rev B 54:R9608
29. Lin H-H, Balents L, Fisher MPA (1997) Phys Rev B 56:6569
30. Matsumoto K, Ogura D, Kuroki K (2018) Phys Rev B 97:014516

Chapter 6
Conclusion

Abstract In this chapter, we will summarize this thesis. We recapitulate main findings concerning electron correlation driven superconductivity in systems coexisting wide and narrow bands, which have been made in the present study. An outlook on future studies is also provided.

Keywords Superconductivity · Electron correlation · Incipient band

In this thesis, we have systematically studied electron correlation driven superconductivity in systems coexisting wide and narrow bands. Specifically, we have addressed two problems concerning a pairing mechanism exploiting an incipient narrow band: (a) to provide a systematic understanding on the role of strong electron correlation effects in this mechanism and (b) to propose a realistic candidate material having coexisting wide and narrow bands with the Fermi level lying close to the edge of the narrow one.

In Chap. 4, superconductivity and strong correlation effects in the bilayer Hubbard model are studied using the multi-band extension of the FLEX+DMFT method, which we have formulated in Chap. 3. By solving the linearized Eliashberg equation numerically, we have shown that a mechanism for spin-fluctuation mediated superconductivity exploiting coexisting wide and narrow bands, which has been mainly discussed within the ordinary FLEX approximation in the previous studies, is, at least qualitatively, still valid even with the vertex corrections coming from the DMFT part of the self-energy. In addition, interestingly, we have found that the present FLEX+DMFT formalism can capture the pseudogap behavior in the spectral function, which can be considered as a hallmark of the strong interband correlation and is hardly seen in the FLEX result. We have also pointed out that there is observed the asymmetric renormalization of the narrow band so that the renormalized density of states is concentrated in the low energy region, which may give an intuitive understanding on the wide applicability of the incipient-band-mediated pairing mechanism among various quasi-one- and two-dimensional systems. Since we neglect intersite vertex corrections in the present formalism, the role of strong correlation effects in a stronger coupling regime, where the Mott transition and related phenomena take place, serves as an interesting future study. We would expect a method combining the FLEX approximation and the cluster extensions of DMFT can be promising.

In Chap. 5, as a possible realistic way to realize high-T_c superconductivity owing to a ladder-type electronic structure, we have introduced a concept of "hidden ladder" in the bilayer Ruddlesden–Popper type layered perovskites. Based on this concept, we have proposed that $Sr_3Mo_2O_7$ and $Sr_3Cr_2O_7$ are candidate materials having an appropriate electron configuration to make the Fermi level sitting in the vicinity of narrow bands from the first-principles band calculation. By performing the many-body analysis for the effective model constructed from the first-principles calculation, we have pointed out that the hidden ladder electronic structure can host high-T_c superconductivity, and we have shown the obtained superconductivity would be essentially the same as that discussed for the two-leg ladder. We have then discussed reasons why superconductivity has not been observed in the actual materials. To realize superconductivity, the material with as little oxygen deficiencies as possible is necessary for $Sr_3Mo_2O_7$. For $Sr_3Cr_2O_7$, an insulating state with the magnetic ordering (and possibly an orbital ordering) should be suppressed. In this case, we have to note that there is a possibility of the Mott transition as well as an orbital ordering, which will be an important issue in the future studies. We have also demonstrated that the present idea of a hidden ladder can be extended to some other materials such as fluorine-intercalated bilayer compounds and trilayer Ruddlesden–Popper compounds, which would provide an interesting future study.

Therefore we would believe that the present study resolve problems in the previous studies concerning electron correlation driven superconductivity in systems coexisting wide and narrow bands at least to a certain extent.

Appendix A
Numerical Analytic Continuation

In this appendix we explain two different numerical analytic continuation methods used in the present study, the Padé approximation and the maximum entropy method. Let us assume we know the numerical value of the Green's function G_n at each Matsubara frequency as

$$G_n = G(i\omega_n). \tag{A.1}$$

Then the problem is how we can obtain the numerical value of the Green's function on the real frequency axis, namely the spectral function. If G_ns are obtained with a sufficient accuracy, one can use the Padé approximation exploiting the continued fraction expansion. On the other hand, if G_n's are obtained with statistical errors typically in Monte Carlo calculations, obtaining the correct spectral function with the Padé approximation becomes difficult. The maximum entropy method is a method to obtain an inference for the spectral functions by picking the one with the highest information-theoretic entropy among possible spectral functions.

Padé Approximation

In the Padé approximation, we approximate as the Green's function taking a continued fraction form:

$$G(\omega) \simeq G_N^{\text{Padé}}(\omega) = \cfrac{a_1}{1 + \cfrac{a_2(\omega - i\omega_1)}{1 + \cfrac{a_3(\omega - i\omega_2)}{\ddots \; 1 + \cfrac{a_N(\omega - i\omega_{N-1})}{1}}}}. \tag{A.2}$$

Here a_i are parameters which are determined through the following procedure, N is the order of the continued fraction expansion, and ω is a complex number corresponding to the frequency. We first define $f_1(\omega_n)$ as

$$f_1(i\omega_n) = G_n, \tag{A.3}$$

and also define $f_j(i\omega_n)$ recursively as

$$f_j(i\omega_n) = \frac{f_{j-1}(i\omega_{j-1}) - f_{j-1}(i\omega_n)}{(i\omega_n - i\omega_{j-1})f_{j-1}(i\omega_n)}, \quad j > 1. \tag{A.4}$$

By this definition, we obtain

$$\begin{aligned}
G_N &= f_1(i\omega_N) \\
&= \cfrac{f_1(i\omega_1)}{1 + \cfrac{i\omega_N - i\omega_1}{1} f_2(i\omega_N)} \\
&= \cdots \\
&= \cfrac{f_1(i\omega_1)}{1 + \cfrac{f_2(i\omega_2)(i\omega_N - i\omega_1)}{1 + \cfrac{f_3(i\omega_3)(i\omega_N - i\omega_2)}{\ddots\ 1 + \cfrac{f_N(i\omega_N)(i\omega_N - i\omega_{N-1})}{1}}}}.
\end{aligned} \tag{A.5}$$

Therefore requiring $G_N = G_N^{\text{Padé}}(i\omega_N)$, a_j can be determined by

$$a_j = f_j(i\omega_j). \tag{A.6}$$

Using the obtained a_i with the above continued fraction expansion, we can calculate $G(\omega)$ for a given ω on the real frequency axis by the following algorithm: First we define P_1 and Q_1 as

$$\begin{aligned} P_0 &= 0,\ Q_0 = 1 \\ P_1 &= a_1,\ Q_1 = 1 \end{aligned} \tag{A.7}$$

and also define P_j and Q_j recursively as

$$\begin{aligned}
P_{j+1}(\omega) &= P_j(\omega) + a_{j+1}(\omega - i\omega_j)P_{j-1}(\omega), \\
Q_{j+1}(\omega) &= Q_j(\omega) + a_{j+1}(\omega - i\omega_j)Q_{j-1}(\omega),
\end{aligned} \tag{A.8}$$

then the Green's function is obtained with

$$G_N^{\text{Padé}}(\omega) = \frac{P_N(\omega)}{Q_N(\omega)}. \tag{A.9}$$

Appendix A: Numerical Analytic Continuation

This relation can be obtained as follows. Using the recursive definition of P_j, we have

$$\begin{aligned}P_j(\omega) &= P_{j-1}(\omega) + a_j(\omega - i\omega_{j-1})P_{j-2}(\omega) \\ &= A_j P_{j-2} + a_{j-1}(\omega - i\omega_{j-2})P_{j-3} \\ &= A_j\left[A_{j-1}P_{j-3} + a_{j-2}(\omega - i\omega_{j-3})P_{j-4}\right] \\ &= \cdots \\ &= A_j A_{j-1}\ldots A_4\left[A_3 P_1 + a_2(\omega - i\omega_{N1})P_0\right] \\ &= A_j A_{j-1}\ldots A_3 a_1,\end{aligned} \quad (A.10)$$

where

$$A_{j-i}(\omega) = \begin{cases} 1 + a_j(\omega - i\omega_{j-1}) & \text{for } i = 0 \\ 1 + \frac{a_{j-i}(\omega - i\omega_{j-i-1})}{A_{j-i+1}} & \text{for } i > 0. \end{cases} \quad (A.11)$$

Similarly for Q_j,

$$Q_j(\omega) = A_j A_{j-1}\ldots A_4\left[A_3 + a_2(\omega - i\omega_1)\right]. \quad (A.12)$$

Therefore

$$\begin{aligned}\frac{P_N}{Q_N} &= \cfrac{a_1}{1 + \cfrac{a_2(\omega - i\omega_1)}{A_3}} \\ &= \cdots \\ &= \cfrac{a_1}{1 + \cfrac{a_2(\omega - i\omega_1)}{1 + \cfrac{a_3(\omega - i\omega_2)}{\ddots\, 1 + \cfrac{a_N(\omega - i\omega_{N-1})}{1}}}} \\ &= G_N^{\text{Padé}}(\omega).\end{aligned} \quad (A.13)$$

Maximum Entropy Method

Let us consider an analytic continuation problem to find the spectral function $A(\omega)$ satisfying

$$G(i\omega_n) = \int_{-\infty}^{\infty} d\omega \frac{A(\omega)}{i\omega_n - \omega}. \quad (A.14)$$

If $G(i\omega_n)$ contains statistical errors, this problem would be ill-conditioned, namely a very small variation in G can give rise to large variations in the resulting A. To avoid this problem, the maximum entropy method employs a different approach, which consists in minimizing the function

$$Q = \chi^2 - \alpha S, \tag{A.15}$$

with

$$\chi^2 = \sum_{mn} (G_m - K_m A)^T C_{mn}^{-1} (G_n - K_n A)^T \tag{A.16}$$

where A is the vector obtained for discretized frequency, K_n is a row vector so that $K_n A$ gives an approximation for Eq. (A.14), and C is the covariance matrix calculated from the statistical samples. S is a relative entropy defined as

$$S = -\int d\omega A(\omega) \ln \frac{A(\omega)}{D(\omega)}, \tag{A.17}$$

where $D(\omega)$ is called the default model and α is a parameter to be determined which is similar to a Lagrange multiplier. $D(\omega)$ is determined so as to include some of the known information in advance about the spectrum. Note that $D(\omega)$ should not have other specific features which may lead to an unphysically biased result.

In the present thesis, we perform the analytic continuation using this method with the ΩMaxEnt code [1].

Appendix B
Convergence of the FLEX+DMFT Results Against Computational Parameters

In this appendix, we will show that the results of the FLEX+DMFT analysis of the bilayer Hubbard model can be considered as being unaffected even if we employ a larger Matsubara frequency range, denser k-point meshes, or a larger cutoff for the Legendre polynomial expansion. Here we employ the wide-and-narrow-band model and take $n = 2.36$, $\beta t = 14$, and $U/t = 5$. Hereafter, we denote the number of k points along each axis as n_k (namely, $n_k \times n_k$ k-meshes) and the number of Matsubara frequencies in total as n_{Mat}, (namely, $n_{\text{Mat}}/2$ positive Matsubara frequencies). Firstly, we show the eigenvalue of the linearized Eliashberg equation for five parameter sets, $(n_k, n_{\text{Mat}}, n_{\text{Leg}})$=(32, 1024, 50), (64, 1024, 50), (32, 2048, 50), (64, 2048, 50), (64, 2048, 70) in Fig. B.1 and Table B.1. As can be seen in the figure, $n_k = 32$ and $n_{\text{Leg}} = 50$ is sufficiently large since the result is unchanged within the statistical errors even if we employ $n_k = 64$ or $n_{\text{Leg}} = 70$.

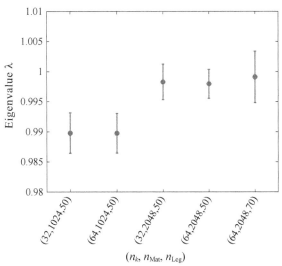

Fig. B.1 The eigenvalue λ of the linearized Eliashberg equation for some sets of computational parameters

© Springer Nature Singapore Pte Ltd. 2019
D. Ogura, *Theoretical Study of Electron Correlation Driven Superconductivity in Systems with Coexisting Wide and Narrow Bands*, Springer Theses,
https://doi.org/10.1007/978-981-15-0667-3

Appendix B: Convergence of the FLEX+DMFT Results Against Computational Parameters

Table B.1 The eigenvalue λ of the linearized Eliashberg equation for some sets of computational parameters

$(n_k, n_{\mathrm{Mat}}, n_{\mathrm{Leg}})$	(32, 1024, 50)	(64, 1024, 50)	(32, 2048, 50)	(64, 2048, 50)	(64, 2048, 70)
λ	0.990 ± 0.003	0.990 ± 0.003	0.998 ± 0.003	0.998 ± 0.002	0.999 ± 0.004

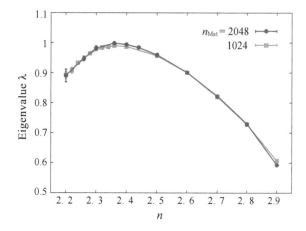

Fig. B.2 The band filling dependence of the eigenvalue λ of the linearized Eliashberg equation for $n_{\mathrm{Mat}} = 2048$. For comparison, the same plot for $n_{\mathrm{Mat}} = 1024$ is also shown

On the other hand, we can see that the result is slightly modified by increasing n_{Mat} from 1024 to 2048. However, since the difference is less than 1%, we would expect this modification does not change the conclusion. To demonstrate this expectation is indeed correct, we show the band-filling dependence of the linearized Eliashberg equation for $(n_k, n_{\mathrm{Mat}}, n_{\mathrm{Leg}}) = (32, 2048, 50)$ in Fig. B.2. It can be seen the result for $n_{\mathrm{Mat}} = 2048$ is almost unchanged from $n_{\mathrm{Mat}} = 1024$ (the same data as in Chap. 4): λ attains a maximum at the same band filling ($n \simeq 2.36$) for both cases and values of λ themselves are quite similar to each other. From this result, we can say that the result of the FLEX+DMFT analysis presented in Chap. 4 can be considered as converged against computational parameters.

Appendix C
Comparison of Spectra Obtained with the Padé Approximation and the Maximum Entropy Method

Here we show that the pseudogap behavior in the spectral functions obtained from the FLEX+DMFT result is unlikely an artifact of the maximum entropy method. We have also performed the analytic continuation of the resulting local Green's function for the anisotropic bilayer model with the Padé approximation. Here we set the band filling as $n = 1.5$. The obtained spectral functions for the bonding/antibonding band is shown in the Fig. C.1. As can be seen from the result, although there are some subtle differences, the overall structure, e.g. a dip in the spectra near the chemical potential and the positions of the peaks, is consistent with each other. This suggests that the obtained pseudogap behavior in the spectral functions is not an artifact of the method.

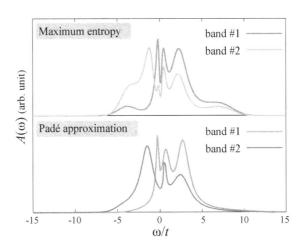

Fig. C.1 Spectral functions for the bonding/antibonding bands in the anisotropic bilayer model. Upper panel depicts those obtained with the maximum entropy method, and lower panel those obtained with the Padé approximation

Reference

1. Bergeron D, Tremblay A-MS (2016) Phys Rev E 94:023303

CPSIA information can be obtained
at www.ICGtesting.com
Printed in the USA
LVHW062321271019
635509LV00004B/10/P